TECHNICAL RESCUER: CONFINED SPACE LEVELS I AND II

George J. Browne
Gus S. Crist

Australia • Brazil • Japan • Korea • Mexico • Singapore • Spain • United Kingdom • United States

Technical Rescuer: Confined Space Levels I and II
George J. Browne and Gus S. Crist

Vice President, Career and Professional Editorial: David Garza

Director of Learning Solutions: Sandy Clark

Product Development Manager: Janet Maker

Managing Editor: Larry Main

Editorial Assistant: Amy Wetsel

Production Director: Wendy Troeger

Production Manager: Mark Bernard

Marketing Director: Deborah Yarnell

Marketing Manager: Erin Coffin

Marketing Coordinator: Shanna Gibbs

Content Project Management: Pre-PressPMG

Art Director: Benj Gleeksman

Compositor: Pre-PressPMG

Cover Designer: Cummings Advertising Art

Cover Image: Stock Studios Photography

© 2010 Delmar, Cengage Learning

ALL RIGHTS RESERVED. No part of this work covered by the copyright herein may be reproduced, transmitted, stored, or used in any form or by any means graphic, electronic, or mechanical, including but not limited to photocopying, recording, scanning, digitizing, taping, Web distribution, information networks, or information storage and retrieval systems, except as permitted under Section 107 or 108 of the 1976 United States Copyright Act, without the prior written permission of the publisher.

For product information and technology assistance, contact us at
Professional & Career Group Customer Support, 1-800-648-7450

For permission to use material from this text or product, submit all requests online at **cengage.com/permissions**
Further permissions questions can be e-mailed to
permissionrequest@cengage.com

Library of Congress Control Number: 2009924857

ISBN-13: 978-1-4283-2410-7

ISBN-10: 1-4283-2410-0

Delmar Learning
5 Maxwell Drive, PO Box 8007
Clifton Park, NY 12065-8007
USA

Cengage Learning products are represented in Canada by Nelson Education, Ltd.

For your lifelong learning solutions, visit **delmar.cengage.com**.

Visit our corporate website at **www.cengage.com**

Notice to the Reader

Publisher does not warrant or guarantee any of the products described herein or perform any independent analysis in connection with any of the product information contained herein. Publisher does not assume, and expressly disclaims, any obligation to obtain and include information other than that provided to it by the manufacturer. The reader is expressly warned to consider and adopt all safety precautions that might be indicated by the activities described herein and to avoid all potential hazards. By following the instructions contained herein, the reader willingly assumes all risks in connection with such instructions. The publisher makes no representations or warranties of any kind, including but not limited to, the warranties of fitness for particular purpose or merchantability, nor are any such representations implied with respect to the material set forth herein, and the publisher takes no responsibility with respect to such material. The publisher shall not be liable for any special, consequential, or exemplary damages resulting, in whole or part, from the readers' use of, or reliance upon, this material.

Printed in United States of America
1 2 3 4 5 6 12 11 10 09

Brief Contents

About the Series vii

About the Authors viii

Preface ix

Chapter 1: Confined Spaces and Their Hazards 1
Chapter 2: Confined Space Entry Requirements 13
Chapter 3: Air Monitoring 30
Chapter 4: Lockout/Tagout 46
Chapter 5: Using the Incident Command System 57
Chapter 6: Strategic Rescue Factors 78
Chapter 7: Ventilation and Inerting 95
Chapter 8: Safety 107
Chapter 9: Rescue 127
Chapter 10: Standard Operating Procedures 154
Chapter 11: Rescue Equipment 164
Chapter 12: Team Evaluation 188

Glossary 203
Index 207

Contents

About the Series   vii
About the Authors   viii
Preface   ix

Chapter 1: Confined Spaces and Their Hazards   1

Lessons Learned: Multiple Deaths in a Confined Space   1
Learning Objectives   2
Introduction   2
Defining Confined Spaces   3
Hazard Recognition   4
   Atmospheric Hazards   4
   Physical Hazards   8
Non-Permit Confined Spaces versus Permit-Required Confined Spaces   9
   Lessons Learned Revisited   *10*
   Summary   *10*
   Review Questions   *11*
   Key Terms   *11*
   Activities   *12*
   Additional Resources   *12*

Chapter 2: Confined Space Entry Requirements   13

Lessons Learned: Workers Die in Construction Accident   13
Learning Objectives   14
Introduction   14
Requirements for Confined Space Entry   15
Confined Space Programs   16
   Attendant   16
   Authorized Entrant   18
   Confined Space Supervisor   18
Confined Space Entry Permit   19
Confined Space Program and Rescuers   26
Permit-Required versus Non-Permit Confined Spaces   26
   Lessons Learned Revisited   *27*
   Summary   *27*

Review Questions   *28*
Key Terms   *28*
Activities   *28*
Additional Resources   *29*

Chapter 3: Air Monitoring   30

Lessons Learned: Two Foremen Die in Rail Car Accident   30
Learning Objectives   31
Introduction   31
Combustible Gases   32
Oxygen Monitoring Equipment   36
Specific Gas Monitoring   37
pH Devices   39
Understanding Monitoring Equipment Readings   40
   SKILLS/PROCEDURES 3-1: *Monitoring a Confined Space*   42
   Lessons Learned Revisited   *43*
   Summary   *44*
   Review Questions   *44*
   Key Terms   *45*
   Activities   *45*
   Additional Resources   *45*

Chapter 4: Lockout/Tagout   46

Lessons Learned: Two Workers Caught in Rupture of Pressure Vessel   46
Learning Objectives   47
Introduction   47
Lockout/Tagout Requirements   49
   Preplanning   49
   Hazard and Risk Assessment   50
   Lockout/Tagout Devices   51
   Lockout/Tagout Equipment   53
   Logout/Tagout Strategic Factors   53
   Lessons Learned Revisited   *54*
   Summary   *55*
   Review Questions   *55*

Key Terms 55
Activities 55
Additional Resources 56

Chapter 5: Using the Incident Command System 57

Lessons Learned: Two Firefighters Trapped in the Collapse of a Burning Building 57
Learning Objectives 58
Introduction 58
Safety 60
Unity of Command 61
Span of Control 61
Common Terminology 62
Single Command and Unified Command 64
The Incident Action Plan 65
Command Post 65
Resource Management 66
Incident Priorities 67
Command 68
   Safety 69
   Public Information Officer 70
Other Command Support Staff 71
   Liaison 71
   Staffing Other Functional Areas 71
   Planning 71
   Operations 71
   Logistics 72
   Finance/Administration 72
Applying the Incident Command System to Confined Space Rescue 73
   Lessons Learned Revisited 75
   Summary 75
   Review Questions 76
   Key Terms 76
   Activities 76
   Additional Resources 77

Chapter 6: Strategic Rescue Factors 78

Lessons Learned: Two Children Trapped in Abandoned Tank 78
Learning Objectives 79
Introduction 79
Basic Rescue Size-Up 80
   Preplanning 80
   Access for Preplanning and Training 80
Basic Strategic Factors 81
   Atmospheric Hazards 81
   Physical Hazards 83
   Exposures 85
   Construction 86
   Contents 87
   Resources 87
   Time 88
   Communication 89
   Identifying the Risk to Life 89
   Weather conditions 90
   Special Problems 91
   Life Safety 91
   Incident Stabilization 92
   Property Conservation 92
   Lessons Learned Revisited 93
   Summary 93
   Review Questions 93
   Key Terms 94
   Activities 94
   Additional Resources 94

Chapter 7: Ventilation and Inerting 95

Lessons Learned: Two Children Trapped in Abandoned Tank, Part II 95
Learning Objectives 96
Introduction 96
Ventilation 96
   Inlet and Exhaust Openings 98
   Power Sources for Ventilation Equipment 101
   Potential Equipment Failures 102
   Positive- and Negative-Pressure Ventilation 102
   Inerting 104
   Lessons Learned Revisited 105
   Summary 105
   Review Questions 105
   Key Terms 106
   Activities 106

Chapter 8: Safety 107

Lessons Learned: Two Rescuers Injured in a Sewage Plant Tank Accident 107
Learning Objectives 108
Introduction 108
Safety Considerations for Personnel 108
   Temperature Stress 110
   Medical Monitoring 113
   Cold-Related Injuries 114
Personal Protective Equipment 116
   Respiratory Protection 116
   Retrieval Equipment 119
   Damage to Chemical Protective Clothing 122
Noise 124
   Lessons Learned Revisited 125
   Summary 125
   Review Questions 125
   Key Terms 126
   Activities 126
   Additional Resources 126

Chapter 9: Rescue 127

Lessons Learned: Contractor Killed and Two Injured in a Lift Station Accident 127
Learning Objectives 128
Introduction 128
Rescue Considerations 128
    Establish command and Take Control of the Incident 128
    SKILLS/PROCEDURES 9-1: *Nine-Step Rescue Process* 129
    Identify the Type of Rescue Problem 132
    Perform a Hazard and Risk Assessment 132
    Identify Rescue Objectives 133
    Identify Resource Needs 134
    Develop an Action Plan 134
    Implement the Action Plan 135
    Evaluate the Effectiveness of the Action Plan 135
    Terminate the Incident 135
Equipment 136
    Tripods 137
    SKILLS/PROCEDURES 9-2: *Managing Three Retrieval/Safety Lines for Two Rescuers and a Victim* 139
    Improvising Lifting Devices 143
    Rope and Equipment 143
    Electrical Equipment 146
    Communications Equipment 147
    Training of Personnel 147
    Termination 148
Initial Scene Operations 148
Accessing the Victim 150
Victim Stabilization 150
Victim Removal 150
    Lessons Learned Revisited *152*
    Summary *152*
    Review Questions *152*
    Key Terms *153*
    Additional Resource *153*

Chapter 10: Standard Operating Procedures 154

Lessons Learned: Two Utility Workers Injured by a Steam Line 154
Learning Objectives 155
Introduction 155
Developing Standard Operating Procedures 155
Written SOPs 156
Checklists 158
    Lessons Learned Revisited *162*
    Summary *162*
    Review Questions *162*
    Key Term *163*
    Activities *163*
    Additional Resource *163*

Chapter 11: Rescue Equipment 164

Lessons Learned: Worker Injured in Fall from a Work Platform 164
Learning Objectives 165
Introduction 165
Types of Loads 166
Equipment Standards 169
Harnesses 170
    Wristlets 174
    Inspection 175
Tripods and Other Legged Rescue Equipment 175
    Lifting Capacity 176
    Surfaces 176
Hoisting Devices and Fall Protection 181
    Ropes and Rope Equipment 183
People and Equipment 185
    Lessons Learned Revisited *186*
    Summary *186*
    Review Questions *186*
    Key Terms *187*
    Activities *187*
    Additional Resources *187*

Chapter 12: Team Evaluation 188

Lessons Learned: Worker Suffers Heart Attack in a Water Storage Tank 188
Learning Objectives 189
Introduction 189
OSHA's Response Time Evaluation 189
    Characterizing the Hazards of a Confined Space 189
    Time Considerations 190
OSHA's Potential Rescue Team Evaluation: Qualifications 191
    Evaluation Components 191
NFPA Standards 194
Other Considerations for Evaluating Confined Spaces 195
Managing Confined Spaces and the Need for Rescue 197
    Rescue Classifications 197
Originally Proposed NIOSH Confined Space Criteria 200
    Lessons Learned Revisited *201*
    Summary *202*
    Review Questions *202*
    Key Terms *202*
    Activities *202*
    Additional Resources *202*

Glossary 203
Index 207

About the Series

NFPA 1006, Standard for Technical Rescuer, 2008 edition addresses the general job performance requirements and specific job performance requirements for special rescue operations. This Technical Rescue series addresses the general job performance requirements and several of the selected special operations portions of the standard. Each of the specialty texts includes both Level I and Level II requirements as defined by the standard. NFPA 1006 defines levels as follows: Level I—This level shall apply to individuals who identify hazards, use equipment, and apply limited techniques specified in this standard to perform technical rescue operations. Level II—This level shall apply to individuals who identify hazards, use equipment, and apply advanced techniques specified in this standard to perform technical rescue operations.

About the Authors

George Browne was appointed to the Weehawken, New Jersey, Fire Department in 1976 as a firefighter. In 1999 he retired as a Chief in the First Battalion with North Hudson Regional Fire and Rescue, the fire department that succeeded the Weehawken Fire Department when it was merged with four other local fire departments in Hudson County, New Jersey. During his career he was the Safety Officer, Training Officer, and Pre-Planning Officer in addition to his line duties. In the mid 1980s George was the Chairman of the Firefighter Health, Safety and Education Committee for the New Jersey State Firemen's Mutual Benevolent Association as Hazwoper, Confined Space Rescue and New Jersey's Public Employees' OSHA came into effect. He has been an EMT and a Haz-mat Technician, and he is trained in Confined Space Rescue. George is a Certified Fire Protection Specialist, a New Jersey State Level II instructor and a certified instructor for Hazardous Materials Technician and Confined Space Rescue. He still teaches for the New Jersey State Police Hazardous Materials Training Unit. Since his retirement, George has been working as a Managing Consultant for Global Risk Consultants and is member of the New Jersey State Industrial Safety Committee and the Technical Committee for NFPA 1620 Pre-Incident Planning.

Gus Crist started his Fire/Rescue service in 1968 with the U.S. Air Force in Crash Rescue. After the USAF he continued in different volunteer organizations including fire, rescue, haz-mat, and emergency management. Prior to his retirement, Gus spent 25 years in the waste water industry and was the Safety Officer for the Ocean County Utilities Authority. During his time with the utilities authority, Gus attended and instructed a variety of courses related to safety and emergency response in the waste water industry. Currently he is an instructor for the New Jersey State Police Hazardous Materials Training Unit and teaches both the haz-mat technician course and the confined space rescue class. Gus was a member of the New Jersey Task Force One steering committee and has been a member of the Task Force since its beginning. He has been appointed to the Ocean County Emergency Council and is a member of the New Jersey State Industrial Safety Committee.

Preface

ORGANIZATION OF THE BOOK

As we wrote this book, we tried to follow the example we used in our original confined space rescue book that was published in 1999: Stick with the basics. Each of the first eight chapters is devoted to an individual topic that you need to know prior to planning any type of confined space rescue operation. If you do not understand confined spaces and their hazards or do not know how to use monitoring equipment or an incident management system, or any of the other basics for confined space rescue, you are operating well beyond your capabilities. Even when you know those basics, you must know how to apply them in a manner that brings them together coherently.

We have tried to incorporate new methods and equipment, and we have devoted the last chapter (which is new) to evaluating confined space rescue teams. There is also a proposal introduced in this book that everyone involved in confined space entry, for whatever reason, must be proactive and look to eliminate hazards. When the hazards can not be eliminated, they must be controlled. Even when you have hazards that are controlled, you must have a plan in place to provide for prompt rescue of the victim if those controls stop working. That might mean attendant-based rescue from outside the space or having a fully trained and equipped team on the site, set up and ready to go to work. And lastly is the idea that as rescuers, we must stop unsafe acts, whether by other rescuers or by the people we may be called on to rescue.

FEATURES

Photos and illustrations

An extensive and up-to-date art program supports the text and focuses on the most current equipment, tools, and procedures.

Learning Objectives

100 percent of NFPA 1006 (2008 Edition) Chapter 7 and the portions of Chapter 3 that relate directly and indirectly to confined space rescue is covered in the Learning Objectives, and other applicable standards in confined space rescue are identified and thoroughly examined.

Lessons Learned

Developed specifically for each chapter, opening vignettes provide real-world insight and application for readers. Each Lessons Learned scenario includes critical thinking questions that tie in to the end of chapter Lessons Learned Revisited section and corresponding discussion questions.

Note and *Safety* Boxes

These boxes highlight important information and remind the reader of important safety precautions and/or procedures.

Skills and Procedures

Step-by-step instructions are combined with photos and line illustrations to show various rope techniques and help the reader gain proficiency.

Review Questions

Chapter-ending review questions test the reader and apply principles in the text materials relating back to NFPA 1006 (2008 Edition).

Activities

Activities are located at the end of most chapters to help readers apply what they have learned. This feature has two purposes: it is designed to assist the instructor with incorporating hands-on practice into the course and give the student skills to practice to ensure mastery of the techniques and concepts presented.

SUPPLEMENTARY MATERIALS

Spend Less Time Planning and More Time Teaching

With Delmar, Cengage Learning's Instructor Resources to Accompany Technical Rescue: Trench, preparing for class and evaluating students has never been easier!

This invaluable instructor CD-ROM allows you anywhere, anytime access to all of your resources:

- The **Instructor's Manual** contains various resources and answers for each chapter of the book and is easily customizable in Microsoft Word.
- The **Computerized Test bank** in ExamView makes generating tests and quizzes a snap. With thousands of questions and different styles to choose from, you can create customized assessments for your students with the click of a button. Add your own unique questions and print rationales for easy class preparation.
- Customizable **PowerPoint® Presentations** focus on key points for each chapter.
- Use the hundreds of *color* images from the **Image Library** to enhance your PowerPoint® Presentations, create test questions or add visuals wherever you need them. These valuable images are pulled from the accompanying textbook, are organized by chapter, and are easily searchable.
- A **Correlation Guide** provides a correlation between *Technical Rescue: Trench* and the requirements for Level I and II, as stipulated by the NPFA 1006 (2008 Edition) standards. These sections from the Standard are correlated to the textbook chapters and are referenced by page numbers.

NOTE FROM THE AUTHORS

There is an old axiom that says that "There are three kinds of people: those who make things happen, those who watch things happen, and those who don't know anything has happened." Both of us are the type of people who like to make things happen. It could be by responding to emergencies, by training other emergency responders, or by developing information or procedures that protect others. Every now and then, if you are lucky, you find out that the thing you made happen saved someone's life or prevented them from serious injury. It is not bravado that drives many emergency responders to do what we do; it is simply the desire to make a difference. Sitting on the sidelines or ignoring what is going on will not suit us. We want to be part of the action. If you are going to make things happen, start by being a professional. Emergency services work is a skilled trade. Learn your trade well.

ACKNOWLEDGMENTS

A book based on technical rescue is, like a rescue, a team effort. No one person could do it alone. Many people allowed us to photograph them during training exercises and graciously gave permission to use their photos. Other people provided information or arranged for equipment demonstrations while still others simply offered sound advice. We would like to recognize the following people for their contributions to this book.

Chief Gregory A. Depaul, Sr., Berkeley Township Emergency Response Team, Bayville, New Jersey

Chief Gregory S. Depaul, Jr., Pinewald Pioneer Technical Rescue Team, Pinewald, New Jersey

Captain Victor Petrucelli, Jersey City Fire Department, Jersey City, New Jersey

Mr. Robert Hansson, Principle Planner, NJ-TF1, New Jersey State Police, Lakehurst, New Jersey

Mr. Marc Brodt, Colgate Palmolive Co., New York, New York

In addition, we would like to thank the following reviewers for their valuable suggestions and insights:

Michael P. Daley, Monroe Township Fire District #3, New Jersey

Steven J. Drozd, Mt. Pleasant Fire Department, South Carolina

Timothy J Ketchmark. Parker Fire Protection District, Colorado

Rick Krumenauer, Oshkosh Fire Department, Wisconsin

George Patterson, Winston Salem Rescue Squad, North Carolina

Dedication

This book is dedicated to all of those who, in doing no more and no less than their sworn duty, have paid the ultimate price to protect others. Whether it was at the World Trade Center, a small single family home on a country road, the scene of a car accident, or a wildfire, these caring men and women thought more of the safety of others than of their own lives.

This book is also dedicated to Dennis L. Devlin, Chief, Ninth Battalion, FDNY.

We also want to dedicate this book to our families. They quietly stood behind us as we worked fires and other emergencies, and they stood behind us as we wrote this book. Thank you to Donna, Helene, Samuel, Sandra, Amy, Dominic, Zachary, Adam, Gussy and Kathryn Cheryl, Timothy and Vanessa, Philip and Jess and Kieran.

NFPA 1006, 2008 Edition, Confined Space Rescue Standard Technical Rescuer: Confined Space, Chapter 7 Correlation Guide

Level 1

Job Performance Requirement	Learning Objective	Chapter Reference	Page Reference
7.1.1	Conduct monitoring of the space	1, 2, 3, 6, 7, 8	5, 14, 42, 81, 91, 96, 108
7.1.2	Prepare for entry	6, 7, 8, 11	80, 97, 108, 185
7.1.3	Enter a confined space	6, 7, 8, 11	82, 97, 109, 185
7.1.4	Prepare the victim for removal	6, 7, 8, 11	84, 96, 109, 180
7.1.5	Remove all entrants and equipment	5, 6, 8, 11	73, 92, 108, 185

Level 2

Job Performance Requirement	Learning Objective	Chapter Reference	Page Reference
7.2.1	Preplan a confined space	4, 6, 8, 9, 10, 11, 12	49, 80, 108, 128, 155, 165
7.2.2	Perform a size up	4, 6, 7, 8, 9, 10, 12	50, 80, 96, 108, 132, 155, 191
7.2.3	Control hazards within the space	4, 6, 7, 8, 9, 10, 11, 12	51, 81, 96, 108, 132, 158, 165, 190

Additional References

Job Performance Requirement	Learning Objective	Chapter Reference	Page Reference
NFPA 1006 3.3.30	Define confined spaces Define permit-required confined spaces	1, 2	3, 16
NFPA 1006 3.3.77	Identify hazardous atmospheres Define flammable atmospheres Describe oxygen-deficient atmospheres Describe oxygen-enriched atmospheres Define toxic atmospheres	1, 6, 7, 9, 10, 12	4, 81, 96, 128, 155, 189
NFPA 1006 5.2	Manage site operations including initial scene operations	5, 9, 10, 12	58, 128, 155, 189
NFPA 1006 5.2.2	Identify the different types of confined space rescue equipment	9, 11, 12	136, 165, 195
NFPA 1006 5.2.2	Categorize operations as defensive or offensive	9, 10, 11, 12	132, 156, 165, 189

Additional References

Job Performance Requirement	Learning Objective	Chapter Reference	Page Reference
NFPA 1006 5.3.3	Recognize the basic victim considerations for assessment, stabilization, and removal	9, 10, 11, 12	133, 155, 165, 189
NFPA 1006 10.1.5	Manage potentially energy sources	4, 8, 9, 10, 12	49, 108, 146, 156, 189
NFPA 1561	Incident Management System	5, 8, 9, 10, 12	58, 108, 128, 155, 189
OSHA 1910.146b	Explain engulfment Define the following basic chemical and physical properties: ■ Flash point ■ Flammable (explosive) range ■ Vapor density	1, 8	4, 108
OSHA 1910.146, Appendix F	Perform an evaluation of a rescue team	12	191
OSHA 1910.147	Lockout/tagout	4, 8, 9, 10	47, 108, 132, 156

1 Confined Spaces and Their Hazards

LESSONS LEARNED: MULTIPLE DEATHS IN A CONFINED SPACE

A father and his two sons decide to spend a Saturday repairing one of the waste tanks on their house's septic system. The father and one of the sons use a ladder to enter the tank and begin working while the other son remains outside to haul out buckets of dry waste. After working for a short time, the son outside the tank notices that his father and brother have collapsed inside the tank. He immediately goes inside the house and calls 911.

Critical Thinking Questions

1. Based on the characteristics of the tank in the above incident, would you identify the tank as a confined space?
2. Would you consider the tank a permit-required confined space or a non–permit-required confined space?
3. If you consider the tank to be a permit-required confined space, what hazards would you anticipate existing or potentially existing in the space?

LEARNING OBJECTIVES

NFPA 1006, *2008 EDITION, OSHA 1910.146, ANSI1-Z117, 2003 Edition*
By the end of this chapter, you should be able to:

- Define confined spaces (NFPA 1006 3.3.30)
- Define permit-required confined spaces (NFPA 1006 3.3.30)
  - Identify hazardous atmospheres
  - Define flammable atmospheres
  - Describe oxygen-deficient atmospheres
  - Describe oxygen-enriched atmospheres
  - Define toxic atmospheres (NFPA 1006 3.3.77)
- Identify mechanical and physical hazards (ANSI Z117.1 3.25, 3.26)
- Explain engulfment (OSHA 1910.146(b)
- Define the following basic chemical and physical properties:
  - Flash point
  - Flammable (explosive) range
  - Vapor density

INTRODUCTION

In the 1980s, the **Occupational Safety and Health Administration (OSHA)** developed a confined space standard and in January 1993 issued the **Code of Federal Regulations (CFR) 29**—Standard 1910.146 entitled *Permit-Required Confined Spaces*. With the development and implementation of this standard and more than 15 years of education and enforcement, you might think that deaths in confined spaces would have become isolated incidents. Unfortunately, they have not.

Just prior to OSHA developing its confined space standard, the **National Institute for Occupational Safety and Health (NIOSH)** studied 8 cases involving deaths that occurred in confined spaces. There were 16 fatalities and 53 injuries; of those 16 fatalities, 10 were would-be rescuers. Most of the time, the rescuers were untrained people who were either part of the confined space work party, or who arrived on the scene after the original entrants were found in the space. As well intentioned as their rescue efforts were, the lack of awareness of the hazards of confined spaces cost these people their lives. As rescue personnel, how many of us might find ourselves in that same position?

The incident in the Lessons Learned is based on an actual confined space incident and was presented in the first edition of this book, published in the late 1990s. It was a typical confined-space accident, with multiple deaths, including would-be rescuers. Now fast forward to 2007. Could a similar scenario happen again?

Consider this scenario: A farmer is attempting to empty an 8-foot deep manure pit on his diary farm when he realizes that the drain pipe is clogged. The farmer enters the pit though a 4-foot wide opening to unclog the pipe. There is about a foot of manure in the pit and shortly after the farmer enters, he collapses. A farm worker, witnessing the farmer collapse, immediately enters to assist the farmer while a second worker notifies the farmer's wife to call for assistance. The farmer's wife runs to the manure pit and enters it to help as the farmer's two young daughters also run to the pit and enter. Emergency crews arriving on the scene find the farmer, his wife, his two daughters, and the farm worker in the pit. And all of them are dead.

This manure pit/confined space was not new to the farm, and the farmer had been cleaning it out about once a week. He may even have successfully entered the pit before to unclog a pipe or clean it out. But what happened this time that led to the death of these people? There was some speculation that a change in the animal feed to include brewer's yeast and a rain storm that washed some of the feed into the pit may have changed the conditions in the space. Was carbon dioxide produced by the brewer's yeast fermenting the manure? Others speculated that methane gas was the cause of the deaths because it is toxic (methane is not toxic). Regardless of what created the atmospheric hazard (was it oxygen-deficient from either the methane or carbon dioxide displacing oxygen, or was there a toxic material present?), five people died. One of those people was the original victim, and the other four were would-be rescuers. Unfortunately, this type of scenario is still common. What would you do differently? Would you be able to recognize the hazards?

> **SAFETY**
> Rescuers must identify the hazards at a confined space incident and protect themselves, and others, from the hazards.

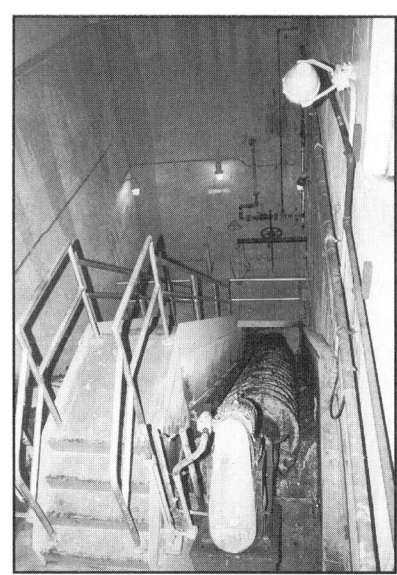

FIGURE 1-1 Interior of a confined space. This space is large enough for a person to enter and work in.

Deaths in confined spaces are terrible tragedies, and they are compounded by the fact that they can be prevented.

DEFINING CONFINED SPACES

What is a confined space and why is it so potentially dangerous? How can we safely operate in a confined space? OSHA's original definition in 1910.146(b) is still applicable today, and it defines a confined space as a space that meets all of the following criteria:

1. Is large enough and so configured that an employee can bodily enter and perform assigned work (Figure 1-1)
2. Has limited or restricted means for entry or exit (for example, tanks, vessels, silos, storage bins, hoppers, vaults, and pits are spaces that may have limited means of entry; Figure 1-2)
3. Is not designed for continuous employee occupancy

These three items are important because they help define the problems with confined spaces, and all three must be present to classify the space as a confined space.

FIGURE 1-2 View of a limited opening on a confined space—opening is on top of the vault.

If the confined space is large enough to enter and to work in, people can enter the space and be directly exposed to any hazards it may contain. Entry initiates the process that exposes the worker to the hazards. Did the septic tank or the manure pit mentioned previously fit the description of the confined space? Did all three criteria exist within those spaces?

Once workers have entered a confined space and come into contact with the hazards, they either have to leave the space on their own or be physically removed by trained rescuers. If the hazard was not obvious, such as an atmospheric hazard, workers entering the space could be overcome by exposure to the contaminated atmosphere and unable to escape.

> **NOTE**
> Confined spaces large enough to enter and to work in allow a person to enter the space and be exposed to any hazards it may contain.

Confined spaces also have limited openings, some which can be smaller than 18 inches in diameter, can be elevated, or can otherwise make entry and exit difficult. Anyone entering the space, including rescuers, must enter through those openings. If the hazards in the space require the use of a **self-contained breathing apparatus (SCBA)** to safely enter and work, it becomes more difficult to enter using those limited or restricted openings. Workers and rescuers may decide to enter the space without using the SCBA, with the intention of donning the equipment after they have entered. Entering the space without SCBA exposes you to whatever atmospheric hazards exist, and you may easily fall victim to the hazards before you are able to don the SCBA. Failing to preplan rescues or poorly developed action plans often lead to this situation. The urgency of the moment during a rescue can blind us to the true nature of the hazards and allow us to become victims.

> **NOTE**
> The sense of urgency many of us feel during rescue work blinds us to the true nature of the hazards and can allow us to become victims.

> **SAFETY**
> Do not remove your SCBA to enter a confined space. Even if you plan on keeping your face piece in place and breathing from the unit, you can drop the cylinder and backpack and have the face piece pulled off your face.

A confined space is not meant for continuous human occupancy. Conditions inside the space can change and no one might be there to notice the changes. Confined spaces that are known to contain hazardous materials or chemical processes are also expected to contain the hazards of those materials, but what about a simple space such as an underground storm sewer? Road runoff, decomposing organic material, and exhaust gases from vehicles can introduce hazards without anyone being aware of their presence.

Ultimately, the problems with confined-space entry for normal work practices, such as maintenance and for rescue operations stem from lack of preparation before entry. There may be unknown and complicated problems. Having an awareness of potential hazards and a plan to work within the limits set by the hazards can lead to a safe operation. Preparation, prior to entry, is a requirement of any confined space entry. Simply because someone has entered a confined space 100 times before without a problem doesn't guarantee entry number 101 will be safe.

> **NOTE**
> Preparation prior to entry is a necessity.

HAZARD RECOGNITION
Atmospheric Hazards

How many times have you been driving down a road and you were able to smell something in the area? It could have been a nearby restaurant, fresh cut grass, or a dump. You probably didn't see a vapor cloud or otherwise become aware of the hazard until you inhaled it. Similarly, the atmosphere within a confined space might be just as invisible. You generally can't tell by looking into a confined space that there is an oxygen-deficient atmosphere or a contaminated atmosphere, and you don't want to be exposed to it by inhaling it. **Atmospheric hazards** in confined spaces are one of the most critical elements affecting confined space operations. These hazards may include toxic gases, flammable gases, oxygen deficiency, or all three. You cannot simply look into the space to determine if an atmospheric hazard exists. You cannot effectively measure the presence of a hazardous atmosphere by sniffing it (that alone can kill you). At one time, coal miners used canaries to detect hazardous atmospheres in the area of the mine they were working in. Since the birds were so small, a smaller dose of coal gas would kill the canary than would kill a person. A dead canary provided an early warning that coal gas was in the area, and the miners could thus escape. Another monitoring device used in the past was a candle, which would be lowered into a confined space. If the candle flame went out that would indicate an oxygen-deficient atmosphere. Imagine what would have happened instead if that space had contained a flammable atmosphere. Fortunately, today's monitoring devices are much more sophisticated, safe, and accurate than canaries or candles.

OSHA Standard 1910.146, Permit-Required Confined Spaces defines a hazardous atmosphere as an atmosphere that may expose employees to the risk of death, incapacitation, impairment of ability to self-rescue (that is escape unaided from a permit space), injury, or acute illness from one or more of the following causes:

1. Flammable gas, vapor, or mist in excess of 10 percent of its lower flammable limit (LFL)
2. Airborne combustible dust at a concentration that meets or exceeds its LFL; Note: this concentration may be approximated as a condition in which the dust obscures vision at a distance of 5 feet (1.52 meters) or less
3. Atmospheric oxygen concentration below 19.5 percent or above 23.5 percent
4. Atmospheric concentration of any substance ... that could result in employee exposure in excess of its dose or permissible exposure limit
5. Any other atmospheric condition that is immediately dangerous to life or health; Note: For air contaminants for which OSHA has not determined a dose or permissible exposure

limit, other sources of information, such as Material Safety Data Sheets (MSDSs)..., published information, and internal documents can provide guidance in establishing acceptable atmospheric conditions.

> **SAFETY**
> Airborne combustible dust at a concentration that meets or exceeds its LFL is explosive when it is in a confined space.

> **NOTE**
> For air contaminants for which OSHA has not determined a dose or permissible exposure limit, other sources of information, such as Material Safety Data Sheets, other published information, and internal documents can provide guidance in establishing acceptable atmospheric conditions.

Flammable atmospheres can be created when gases, liquids, dusts, and vapors are present in a confined space. Normal process operations and work being performed in the space or accidental leaks or spills are just some examples of how these materials can be introduced into the space.

In understanding the hazards of a flammable atmosphere within a confined space, it is important to understand certain basic principles about the fuel.

> **NOTE**
> Sufficient fuel, oxygen, and heat must be present for combustion to occur.

The flammability of these materials differs from material to material. For these materials to ignite, there must be a minimum concentration of the materials present, an ignition source with enough energy to ignite the material, and sufficient oxygen for combustion to occur. When working around fuel vapors (from gases, certain liquids, and solids), you must understand the **flammable** or **explosive range,** as shown in Figure 1-3. The flammable range has a lower and an upper flammable limit (also called lower

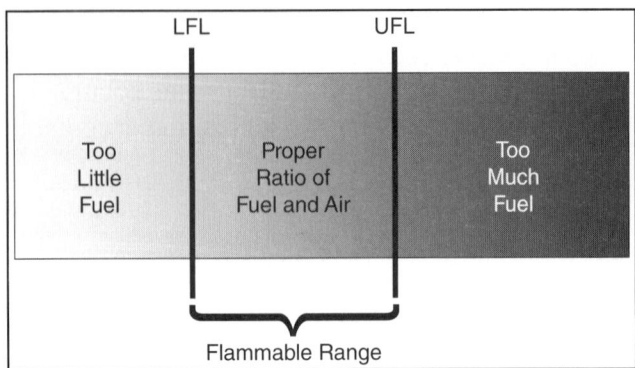

FIGURE 1-3 Illustration showing flammable range. The fuel mixture to the left of the LFL is too lean (not enough fuel), and the fuel mixture to the right of the UFL is too rich (too much fuel, not enough oxygen).

explosive and upper explosive limits). These limits are the minimum (lower flammable limit) and maximum (upper flammable limit) concentrations of the fuel in air needed to support combustion. If the mixture is below the lower flammable limit (LFL), it is considered too lean to burn. If the mixture is above the upper flammable limit (UFL), it is considered too rich to burn. In addition, if the temperature of the fuel and air mixture is increased, the flammable range will also increase, whereas if the temperature is decreased, the flammable range will also decrease.

> **NOTE**
> NFPA and OSHA use LFL and UFL, whereas the fire service often uses LEL and UEL. For the purposes of this book, these terms are interchangeable.

In discussing flammable limits, we must also recognize that certain liquids produce enough vapors at normal temperatures and atmospheric pressures that the presence of these liquids within a confined space can create a flammable mixture. Typically these flammable liquids will have a **flash point,** as shown in Figure 1-4, at or below the ambient temperature within the space. The flash point of a material is the temperature at which an ignitable mixture is created. Flash points are determined in a laboratory and can vary slightly depending on the method used. A good rule to follow is that when the temperature

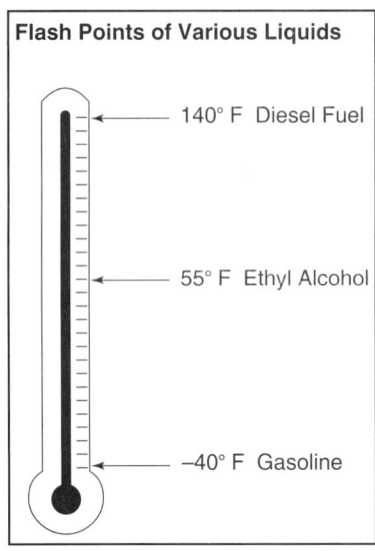

FIGURE 1-4 Different materials have different flash points. Knowing the identity of the material, the flash point, and the temperature of the material will give you an idea of the fire hazard.

of a liquid is at or above its flash point, it should be considered at or above its lower flammable limit.

Flammable atmospheres from solids generally become a problem when the solids are present as dust. Dusts are minute particles of a solid. Because these particles are so finely divided, they are easily heated by an ignition source. Once heated, these dusts ignite and burn. When contained in a confined space, this flash fire creates a rapid buildup of pressure and an accompanying pressure wave. This is a dust explosion.

> **NOTE**
> A flash fire in a confined space is an explosion and creates a pressure wave because it is contained. Depending on the force of this pressure wave, it can injure and kill people and destroy property.

To categorize the hazard presented by the presence of a combustible dust, OSHA uses as an *approximation* of the condition when the dust obscures vision at a distance of 5 feet (1.52 meters) or less. Although this is an approximation, the presence of a dust condition of this magnitude should be cause for immediate concern.

To help you better understand dusts and their combustibility, here is some basic information about combustible dusts:

- The dust particles must be finely divided and be about 420 microns or smaller (about the size of table salt or sugar particles).
- The smaller the dust particles, the easier they are to ignite and, typically, the larger the explosion.
- A dust explosion may not be a single event. When the dust ignites, the pressure wave can dislodge other dust in the area, causing a second, third, or more explosions.
- As a rule of thumb, if combustible dusts are present and the layer of dust is deep enough that you can write your name in the dust, you have enough dust to create an explosion.

Not all materials have the same flammable range. Gasoline has a given flammable range of 1.4 to 7.6 percent in air, while anhydrous ammonia has a flammable range of 15.5 to 27 percent in air and acetylene has a flammable range of 2.5 to 100 percent in air under certain conditions, as shown in Figure 1-5. As the flammable range of the materials varies, so does the potential hazard presented by the material. A 2 percent concentration of gasoline in air is within the flammable range and with an appropriate ignition source, you can expect a fire. A 2 percent concentration of either anhydrous ammonia in air or acetylene in air is not within the flammable range.

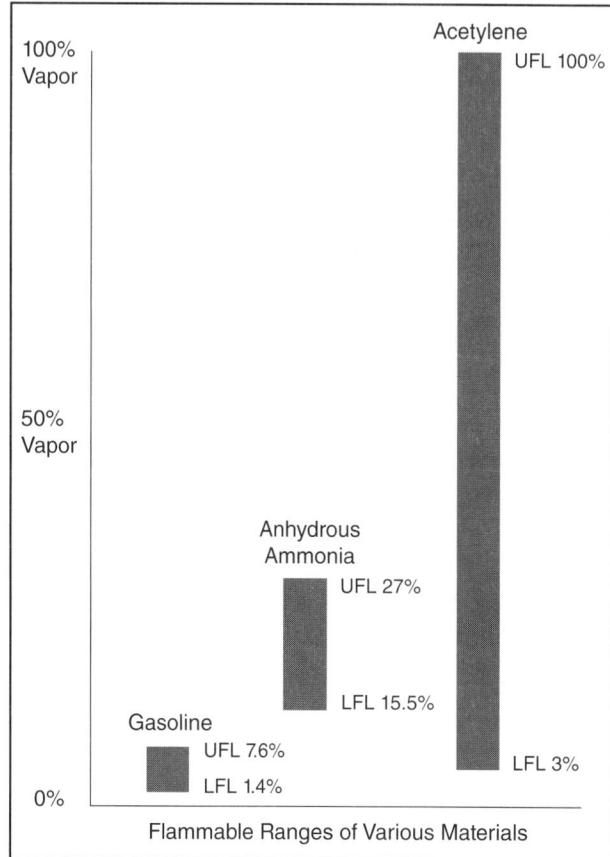

FIGURE 1-5 Flammable ranges of various materials. Some materials have a narrow flammable range, whereas others have a broad range.

> **NOTE**
>
> It is important to realize that, as the flammable range of the materials varies, so does the potential hazard presented by the material.

OSHA uses a limit of 10 percent of the LFL to accommodate the variations in materials. Observing a 10 percent LFL action limit creates a greater margin of safety in that most metering equipment for measuring combustible gases is **calibrated** on a single gas. The calibration gas has a specific known LFL. Staying below 10 percent of the LFL accommodates other gases that may be present and have an LFL lower than the gas used for calibrating the meter.

As discussed previously, there must be adequate oxygen for combustion to occur and, more importantly, for the support of human life within the confined space. The atmospheric air we breathe normally contains approximately 21 percent oxygen. When the concentration goes below 19.5 percent, the air is considered oxygen-deficient and poses a danger to human life, as shown in Figure 1-6. Low oxygen concentrations can also affect the flammability of fuels in the space. A potentially explosive atmosphere can be present in the space, but the fuel air mixture lacks enough oxygen to ignite. An increase in oxygen can raise the mixture into the explosive range and bring together all ingredients needed for a fire or explosion. An oxygen-deficient atmosphere is not the only oxygen-related problem in a confined space. Too much oxygen creates an oxygen-enriched atmosphere and when the oxygen concentration is above 23.5 percent, other hazards are brought in. Oxygen-enriched atmospheres rapidly

Effects of Reduced O₂	
21%	Normal atmosphere
19.5%	OSHA definition as oxygen-deficient
17%	Some muscular impairment, increased respiratory rate
12%	Dizziness, headache, rapid fatigue
9%	Unconsciousness
6%	Death within a few minutes

FIGURE 1-6 Effects of varying levels of oxygen on people.

accelerate combustion, and materials that are not easily ignited in normal air or are not normally considered combustible can be ignited and burn in an oxygen-rich atmosphere. The oxygen levels within the confined space can be affected by what occurred in the space prior to entry, by the type of work being done, or even by sources outside of the space.

SAFETY

Oxygen concentrations below 19.5 percent are considered oxygen-deficient, pose a danger to human life, and are one of the most common causes of confined space emergencies.

NOTE

Oxygen concentrations above 23.5 percent are considered oxygen-enriched.

Any discussion of atmospheric hazards of confined spaces would not be complete without considering the presence of hazardous materials. The presence of hazardous materials can be created by many different means. The simplest source of hazardous materials is the products stored or processed in the confined space. Other sources include the materials brought into the area by those performing work there. Still other sources might be chemical reactions occurring in the space such as the decomposition of organic material either in the area or in the ground surrounding it.

SAFETY

Hazardous materials can create toxic atmospheres that can injure or kill victims and rescuers through exposure to the materials (inhalation, ingestion, or direct skin contact), displacement of oxygen, or a combination of the two.

It is plain to see why atmospheric hazards alone can create life-threatening conditions within a confined space, and it is just as easy to see why an unprepared rescuer can become a victim. Firefighters would not start fire operations without sizing up the situation, and EMTs would not start treatment on their patients without assessing them first. Why as a rescuer at a confined space accident would you want to do any less? If you saw a person drowning, what would you do to rescue them? Would you jump into the water and attempt to swim to the victim, or would you call for help and then find other equipment such as a boat or a rope to help him? In a confined space incident, your victim could be drowning in an atmosphere that is toxic, oxygen-deficient, or explosive. What resources do you need to complete a safe and successful rescue?

Physical Hazards

We have spent some time discussing atmospheric hazards, but not all hazards you will face in a confined space will be atmospheric hazards. **Physical hazards** can also be present and must also be considered in your size-up. Physical hazards include electrical equipment, agitators or paddles (shown in Figure 1-7), fire suppression equipment, sharp edges (shown in Figure 1-8), difficult-to-access areas due to height, and the physical state of the material present in the space, as well as other considerations. **Engulfment** by the materials within the space is a common physical hazard. A liquid or finely divided (flowable) solid substance can surround and trap a person. When a victim is engulfed by a product, such as sawdust, she can aspirate the material and die due to the respiratory system being filled or plugged. Other materials can flow with gravity, engulf a person, and exert enough force on the body to cause death by strangulation, constriction, or crushing.

FIGURE 1-7 A variety of physical hazards can be present in a confined space. Note the sludge thickener (screw) running up the center of the picture.

Do not fail to consider physical hazards as you assess the conditions presented by a particular confined space emergency. Not only do these hazards endanger the victim, they must be controlled so that they do not endanger the rescuers. Among the methods to control physical hazards is a procedure known as lockout/tagout. Chapter 4 in this book addresses lockout/tagout.

FIGURE 1-8 This is a final clarifier tank in a sewage treatment plant. The hazards shown here include the water, sloped sides near the top of the tank, and the weirs (pointed metal triangles).

NON-PERMIT CONFINED SPACES VERSUS PERMIT-REQUIRED CONFINED SPACES

So far our discussion has been about confined space in general. All confined spaces meet the following three criteria: large enough and so configured that an employee can bodily enter and work, limited or restricted means for entry or exit, and not designed for continuous employee occupancy. But not all confined spaces contain hazards. There is a difference between a confined space and a permit-required confined space. OSHA has classified confined spaces as **permit-required confined spaces** and **non-permit-required confined spaces** to better differentiate between confined spaces that are expected to contain hazards to people and those that do not. Non-permit confined spaces are spaces *that do not contain or, with respect to atmospheric hazards, have the potential to contain any hazard capable of causing death or serious physical harm.*

A permit-required confined space is an area that meets the three criteria for confined spaces and that has one or more of the following characteristics:

1. Contains or has the potential to contain a hazardous atmosphere
2. Contains a material that has the potential for engulfing an entrant
3. Has an internal configuration such that an entrant could be trapped or asphyxiated by inwardly converging walls or by a floor that slopes downward and tapers to a smaller cross section
4. Contains any other recognized serious safety or health hazard

As rescuers, you must be aware of the difference between permit-required and non-permit confined spaces. A permit-required confined space requires different precautions to protect workers in the space than a non-permit required confined space does. The requirements to protect workers in a permit-required space may include ventilation, respiratory protective equipment, personal protective clothing, monitoring equipment, and retrieval equipment. Because you might not expect to find this equipment at a non–permit-required space, if you arrive at a confined space accident, are told that the space was initially classified as a non-permit space, and do not see any of this equipment present, you may be tempted to consider that you are not about to embark upon a confined-space rescue. Your initial impression does not necessarily indicate that the space contains no hazards. Depending on the type of work being performed in the space, new or unknown hazards may have been introduced into it. These hazards can easily turn that non-permit space into one with serious hazards—hazards so serious, in fact, that the space should be reclassified as a permit-required space. Further complicating matters is the fact that the OSHA general industry standard 1910.146 does not apply to agriculture, construction, or shipyard employment (for shipyards, reference CFR 29—1915 Subpart B, *Confined and Enclosed Spaces and Other Dangerous Atmospheres in Shipyard Employment*). These sites may have confined spaces present, but you may have to identify them as such or identify the hazards unique to those sites (CFR 1915.11 uses a limit of 22.0 percent oxygen by volume to define oxygen-enriched atmospheres because of **hot work**). As a rescuer, you should treat all confined spaces as permit-required spaces. Only after the space has been classified by monitoring, assessing, and reviewing the problem should you consider reclassifying it to a non-permit space. If you have any doubt as to the need for a permit, treat the space as a permit-required confined space.

> **NOTE**
>
> Rescuers must be aware of the difference between permit-required and non-permit confined spaces. Permit-required spaces contain or have the potential to contain serious safety and health hazards including hazardous atmospheres, engulfment or an internal configuration that could trap or asphyxiate a person.

> **NOTE**
>
> It is in your best interest to treat all confined spaces as permit-required spaces.

LESSONS LEARNED REVISITED

After calling 911, the son returns to the septic tank and enters the tank opening to assist his father and brother. In the end, the father and both sons die before they can be rescued.

Discussion Questions

1. Had you been a witness to this incident and seen the victims prior to the arrival of trained rescuers, what would your initial reaction have been?
2. Based on how you answered question 1, do you think your initial reaction would be correct or incorrect?
3. If you had to make entry into the tank, name at least three precautions you would take prior to entry.

SUMMARY

- Confined space emergencies can and have killed workers and rescuers.
- Recognizing confined space emergencies and knowing the dangers can help you to protect yourself as you attempt to rescue victims.
- Recognizing the presence of a confined space is the first step to aiding victims.
- Confined spaces meet all of the following criteria:
  - Large enough to enter
  - Have limited means for entry or exit
  - Are not designed for continuous human occupancy
- You must identify potential atmospheric and physical hazards in the space.

- You must define the effects these hazards will have on your rescue operations.
- The strategic factors presented by the conditions in and around the space can dramatically change your method of operation.
- You must recognize when to call for additional help.

REVIEW QUESTIONS

1. What three separate items define a space as a confined space?
2. If you respond to an emergency that has occurred in what initially appears to be a confined space, but your size-up of the situation shows that only one of the confined space conditions exists, should you treat the space as a true confined space?
3. Are the flammable range and lower and upper limits identical for all materials?
4. What are the acceptable oxygen concentrations within a confined space?
5. Identify two potential physical hazards within a confined space.
6. The vapor density of air is given as 1. Is a gas that has a vapor density of 3.1 lighter or heavier than air?
7. Using the sense of smell to detect the presence of hazardous gases or vapors is a recommended practice. True or False?
8. For combustion to occur, there must be _____.
    a. fuel, heat, and oxygen
    b. sufficient fuel, heat, and oxygen
    c. fuel, heat, and an uninhibited chemical reaction
    d. heat, an uninhibited chemical reaction, and oxygen
9. The process by which a liquid or finely divided solid substance surrounds and traps a person is known as _____.
    a. asphyxiation
    b. strangulation
    c. engulfment
    d. physical hazard
10. The flammable range for gasoline is _____, and the flammable range for acetylene is _____.
    a. 1.4 to 7.6 percent in air, 15.5 to 27 percent in air
    b. 2.5 to 100 percent in air, 1.4 to 7.6 percent in air
    c. 2.5 to 100 percent in air, 15.5 to 27 percent in air
    d. 1.4 to 7.6 percent in air, 2.5 to 100 percent in air

KEY TERMS

Occupational Safety and Health Administration (OSHA) 2
Code of Federal Regulations (CFR) 2
National Institute for Occupational Safety and Health (NIOSH) 2
Self-contained breathing apparatus (SCBA) 4
Atmospheric hazards 5
OSHA Standard 1910.146, Permit-Required Confined Spaces 5
flammable or explosive range 5
flash point 6
calibrated 7
physical hazards 8
engulfment 8
permit-required confined spaces 9
non-permit confined spaces 9
hot work 10

ACTIVITIES

1. Download additional Fatality Assessment and Control Evaluation (FACE) Program reports on confined space accidents from the NIOSH website and discuss the various hazards associated with the accidents.

2. Go to the OSHA website and under the section titled Laws and Regulations, review the official interpretations of the OSHA Confined Space Standard 1910.146.

ADDITIONAL RESOURCES

http://www.cdc.gov/niosh/face/

http://www.osha.gov/

2 Confined Space Entry Requirements

LESSONS LEARNED: WORKERS DIE IN CONSTRUCTION ACCIDENT

Two workers were killed at a construction site when they were overcome by fumes in a pit being dug for a new sewage pumping station. Police and EMS were called to the site after a supervisor saw the two workers lying in water at the base of the pit and called to report that two people had fallen off a ladder. The pit was 15 feet square and 27 feet deep. The pumping station was being built at the edge of the marshes near the end of Main Street, and the area has a high water table.

Critical Thinking Questions

1. Based on the characteristics of the construction site in the preceding incident, would you identify this construction pit as a confined space?
2. Would you consider the pit a permit-required confined space?
3. If you consider the pit to be a permit-required confined space, what hazards could potentially exist in the space?

LEARNING OBJECTIVES

NFPA 1006, *2008 EDITION*
By the end of this chapter, you should be able to:

- Understand the purpose and the basis of a confined-space entry program (3.3.30)
- Identify the functions of the attendant, authorized entrant, confined-space supervisor
- Define the following (3.3.34):
  - Permit-required confined space
  - Non-permit confined space
  - Confined space permit

INTRODUCTION

OSHA has estimated that more than 4.8 million permit-required confined spaces are entered annually. That means employers must manage hose millions of entries and provide for the safety of the workers who are assigned to work in the permit-required spaces. Of those millions of entries, how many are made safely using a safe procedure and following basic precautions? How many near misses occur, and how many times do the workers in the space not recognize the potential danger? How many accidents with injuries and deaths occur?

> **NOTE**
> Rescuers should treat all confined spaces as potentially hazardous until a thorough evaluation of the conditions shows that no hazards exist.

Given the information in Chapter 1, you know that confined space accidents can be prevented. In 1977, the American National Standards Institute (ANSI) first published Z117.1-2003, *Safety Requirements for Confined Spaces*. This voluntary (but nationally recognized) standard was the model that some states and many private companies used to develop their own programs for confined space entry prior to OSHA's adoption of 29 CFR 1910.146. With the use of a voluntary standard, these programs varied from simple generic programs to site-specific, detailed programs. Some trade organizations also recognized that their industries (for example, refineries, bulk storage terminals) shared an industry-wide hazard, and these trade organizations also developed voluntary, model programs. Regardless of the source, the goal of these programs was still the same, to provide for worker safety during entry into confined spaces.

One such program was used in the brewing industry. After the brewing tanks were emptied, workers were required to enter and clean the tanks. Beer is meant for consumption, and the only atmospheric hazard that remained in the tank after brewing was any remaining carbon dioxide. The brewing tanks would be emptied, ventilated, and then tested for oxygen content by lowering a miner's lantern into the tank. Depending on the oxygen content of the tank, the lantern would continue to burn, indicating the oxygen level was adequate to support human life or, if the lantern extinguished, the atmosphere was oxygen-deficient. As primitive as this procedure may sound, it was an effective method until better means became available to test the atmosphere in the tank. The problem, however, was that not all employers had gone through confined space programs. This inconsistency and the resulting injuries and loss of life between programs led to the development of the OSHA Confined-Space Standard, CFR 1910.146, described in Chapter 1.

REQUIREMENTS FOR CONFINED SPACE ENTRY

The OSHA confined space standard has many requirements that range from training employees to recognize a confined space, differentiating between permit-required and non-permit required space, precautions to take prior to entry, to the requirements to be followed during entry. As rescuers, we need to be aware of all of the requirements of 1910.146, including that there be a means to rescue people from the space in the event of an accident. This is where your involvement as a rescuer begins—with safe entry into the space (as shown in Figure 2-1) to help people who might have been injured or trapped by an accident. Simply because you are at the incident as a rescuer does not exempt you from exposure to the same hazards and injuries as the victims. Prior to your entry, you must take precautions to manage the risks presented by the confined space. The intent of a confined space program is to prevent accidents, so if you are called to rescue people at a confined space incident, you'll know that something has gone wrong with this program. This "something" can be as simple as the employer not providing a confined space program or the accidental introduction of a hazard into the space after such a program was initiated. To determine what has happened, knowing the requirements of a confined space program and comparing it with the current situation will assist you in determining what created the emergency. You, as a rescuer, must also be know the requirements of a confined space entry program so that your agency can create standard operating procedures (SOPs) to protect you during the emergency. Without confined space rescue SOPs and operational guidelines to follow, the only difference between the victim and the rescuer is that the rescuer hasn't entered the space yet. If you fail to recognize the hazards, take precautions to deal with them, and provide for the potential rescue of members of your team working in the space, you run the risk of becoming a statistic.

FIGURE 2-1 Confined space entry permit being filled out prior to entry into a confined space.

> **NOTE**
> More than 60 percent of the fatalities in confined space accidents are untrained rescuers.

With that said, let's look at the requirements for confined space entry. As you examine these requirements, it should become obvious that most of the items in OSHA's entry program, as well as those of ANSI and other nationally recognized standards, provide valuable information for sizing up an incident. An **entry permit** is a written document that outlines what will occur within the space, including the type of work, the expected hazards, and required precautions. When you do not have an entry permit, where do you start your evaluation of the incident? You will have to spend time determining what has happened, how you can immediately assist the victim (defensively), and what steps you must take to protect yourself as you prepare to enter and effect the rescue (or recovery). Even when a permit is present, you must verify that the information on the permit is correct. Knowing that attendants, entrants, and an entry supervisor are required and must be identified on the permit will allow you to determine who you should be talking with to find out what happened. How can you account for the workers at the scene without interviewing the people with the most knowledge about the situation? Consider the following:

- Get a head count and identify who is missing.
- Are there more victims than first believed?

- Did untrained rescuers enter the space, and have they added to the victim count?
- What work was being done?
- What were the expected hazards, and what happened at the time someone realized there was an emergency?

These are all questions you need to get answers to as quickly and accurately as possible. So who do you talk to? The entry permit should be where you begin, and creating a rescue entry permit (not required, but a good idea) modeled on OSHA's model permit is where you should begin.

CONFINED SPACE PROGRAMS

Employers must begin their confined space program by evaluating the workplace to determine if any permit-required confined spaces exist. *If the workplace contains permit-required spaces, the employer shall inform exposed employees, by posting danger signs or by any other equally effective means, of the existence and location of and the danger posed by the permit spaces 1910.146(c)(2).* These signs may say "Danger—Permit-Required Confined Space. Do Not Enter" or other language to effectively communicate the presence of a permit-required confined space (see Figure 2-2). But what if the evaluation reveals only non-permit confined spaces? You will not find signs posted in such areas. Should you interpret that to mean that those non-permit spaces don't pose a hazard? No. As a rescuer, you should treat all confined spaces as potentially hazardous until a thorough evaluation of the conditions shows that no hazards exist.

> **NOTE**
> A confined space program can minimize the effects of an accident and speed your rescue operations.

In some cases, an employer may have decided that its employees will not enter confined spaces, leaving work requiring confined space entry to contractors. If this is the case but you have been called to make a rescue at this site, how did the victim get into the confined space? Was this entry in violation of the company policy, or is the victim an outside contractor working at the site? Regardless of why you have been called to the incident, as part of the rescue team you must take control of the situation. Who do you talk to, how do you evaluate the incident—where do you start?

To simplify the process of creating rescue procedures for confined spaces, you should begin by looking at the relevant OSHA requirements. A key place to start is with the roles and responsibilities that people are given during a permit-required confined space entry. Although the roles were originally intended for workers entering the space, these same roles can be adapted for rescue purposes and should be used to identify competencies and provide for the safety of rescue team members. There is no need to reinvent the wheel and to simplify the process, the roles and responsibilities identified in this chapter will only be those required by OSHA for workers who will enter and work in confined spaces. As rescuers, your team's roles and responsibilities will closely parallel those of the workers, but as you work your way through this book, the differences will become clearer.

Attendant

Who is the attendant? What does she do? And how does this work add to the safety of the rescue team? The OSHA standard (1910.146) defines the duties of and the requirement for an **attendant**. An attendant is an individual who is stationed outside one or more permit-required

FIGURE 2-2 As part of a confined space entry program, confined spaces are required to be identified.

Confined Space Entry Requirements 17

FIGURE 2-3 Attendant communicating with the entrants working inside the confined space.

FIGURE 2-4 Tools and equipment lying around the outside of a confined space present a danger because they can fall or be kicked into the space.

confined spaces and monitors the people within and the conditions in the area around the space, as shown in Figure 2-3. The attendant may not enter the space for any reason. She may not bring tools or equipment into it, perform rescue by entering the confined space, leave the area, or enter the space for any other reason unless she is relieved by another trained attendant.

> **NOTE**
> The attendant's role is critical because that person is the outside link for those in the confined space.

Attendants are also responsible for monitoring the people within the space and providing **Standby Assistance** (ANSI Z117.1, 2003). They ensure the safety of the people within the space and maintain communication with the workers in the space. Attendants can call for help if an accident occurs and can order the space evacuated if conditions either within or outside of the space pose a hazard to any person.

The critical nature of the attendant's role is that he is the outside link for workers in the confined space. The attendant not only monitors what is going on inside the space (atmospheric hazards, workers behavior, physical condition, and so on), but he also watches for hazards outside. Without an attendant, how would the workers in the space know that a fire alarm was sounding in the building where the confined space was located? Without an attendant, who would be responsible for ensuring that the area around the opening to the space is clear of tools and equipment, as shown in Figure 2-4? These tools could easily be kicked or knocked into the space and onto any workers there. If you are called to a confined space emergency, that attendant may be the single most valuable source of information you have. This person witnessed the incident and may also have made initial rescue attempts. The attendant is required to remain alert to the situation and may feel responsible for the accident. When you are interviewing the attendant, be aware that he may feel responsible for the accident and provide less than complete information.

Attendants can legitimately attempt a rescue from outside of the confined space. In that case, the employer must develop non-entry rescue procedures to be followed by attendants, provide appropriate rescue equipment (for example, tripods, winches), and properly train attendants to use the procedures and equipment. This process is called **attendant-based rescue**.

> **NOTE**
> The attendant is not to enter the space to bring tools or equipment into it, to perform rescue, or for any other reason unless relieved by another trained attendant.

As part of the procedures, roles, and responsibilities you develop for your confined space rescue team, you should also identify the functional position of attendant as part of your team for any confined space rescue. The attendant is identified as a **rescue attendant** and carries out all of the attendant duties for the rescue team.

Authorized Entrant

The purpose of any confined space program is either to make the area safe for people to enter or to keep everyone out. As shown in Figure 2-5, the people who will enter the space to complete an assigned job have roles and responsibilities. When these entrants are trained to be aware of the hazards, controls, roles, responsibilities, needed equipment, and other requirements for safe entry, they will become **authorized entrants**. Essentially, the authorized entrants must be informed of the hazards that exist or can occur in the confined space, must know how to use required personal protective equipment, must recognize any warning signs or symptoms of a dangerous condition, must alert the attendant as to when a dangerous condition is detected, and must exit from the space as quickly as possible when necessary.

Typically the entrant is at least one of the victims at a confined space accident. The entrant is in the space and exposed to hazards, and may not be able to leave the space without the assistance of an outside rescuer. Because authorized entrants are expected to be trained to enter the space safely, if they are victims, you should immediately investigate the situation to determine what happened.

- Were they authorized to enter the space?
- Did some piece of equipment fail and change the conditions in the space?
- Was there a change in the materials to be used or the work being done in the space?

Remember, not all of the people in a confined space will be authorized entrants. Find out if the victims are authorized entrants or if they are unqualified people who did not know the hazards they faced.

Just as you should develop procedures, roles, and responsibilities for a rescue attendant as a part of your confined space rescue team, you should do the same with the functional position of authorized **rescue entrant**. The rescue entrant will carry out duties that parallel those of the authorized entrant in the OSHA Confined Space Standard except that they enter for the purposes of rescuing victims from the confined space.

Confined Space Supervisor

Considering the requirements of a confined space entry program, someone must be placed in charge of the program during permit-required entries. OSHA identifies this person as the **confined-space supervisor.** This supervisor has a variety of tasks and obligations to fulfill. The supervisor is required to:

- Know the hazards that may be faced during entry, including the mode, signs or symptoms, and consequences of the exposure. They must inform the attendant and entrant of the hazards.
- Verify, by checking that the permit has been filled out properly, that all tests specified by the permit have been conducted, and that all procedures and equipment required by the permit are in place before allowing entry to begin, as shown in Figure 2-6.
- Terminate the entry and cancel the permit as required.
- Verify that rescue services are available and that there is a means for summoning them.

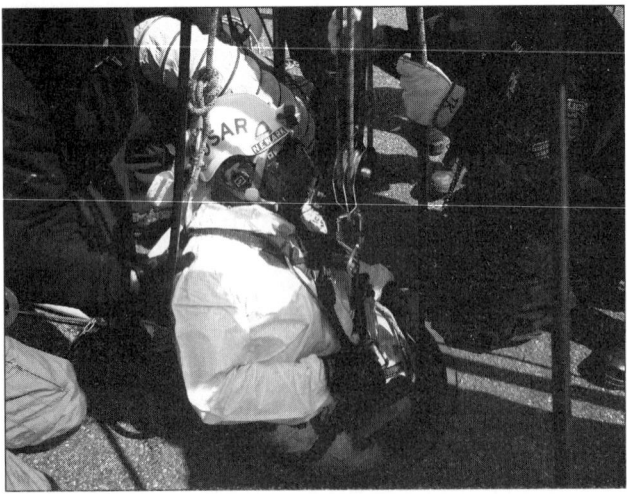

FIGURE 2-5 Any person who makes entry into a confined space is an entrant and must be trained as an entrant. This entrant training requirement includes rescuers.

FIGURE 2-6 The confined space supervisor is responsible for making sure that the entry permit is properly filled out.

- Remove unauthorized individuals who enter or who attempt to enter the permit space.
- Determine that entry operations remain consistent with the terms of the entry permit and that acceptable entry conditions are maintained at all times.

The entry supervisor should be the most well-informed person on the scene of a confined space accident, but to be so well-informed, there must be a confined space entry program in place. As discussed earlier in this chapter, the intent of such a program is to prevent accidents, but if an accident occurs, who or what is the cause? Was a hazard introduced into the space that caught the workers off guard, or is there no confined space entry program in place? With no program there is no entry supervisor, no trained entrants, and no trained attendant. Without trained people at the scene, how do you get accurate information to evaluate what has happened and then develop an action plan?

Lacking accurate information, you will have to begin at the most basic level to protect yourself and other emergency responders. Your basic goal should be to prevent accidents during the rescue, minimize the impact of potential accidents, and efficiently complete your rescue operations. The information you will need to make decisions during a rescue is only as good as the source of that information.

In addition to the rescue attendant and rescue entrant, you will need to have someone to supervise the rescue operation. Though this functional role might easily be performed by the incident commander or the operations officer, they must be qualified to take on this supervisory role. The qualifications for the person who is to be the rescue supervisor are similar to the role defined by OSHA for Confined Space Supervisor, but the rescue supervisor must comprehend the roles and responsibilities of the rescue attendant and rescue entrants, the limitations of the rescue team's procedures and equipment, the OSHA confined space standard, and their position within the incident command system.

CONFINED SPACE ENTRY PERMIT

So far, we have discussed confined spaces and permit-required confined spaces. Those two types of spaces are both addressed in the OSHA standard. The ANSI Standard titled *Safety Requirements for Confined Spaces* (Z117.1-2003), though still defining a confined space in terms identical to those used by OSHA, better differentiates confined spaces as either *non-permit required confined spaces (NPCS)* or *permit-required confined spaces (RPCS)*, as previously discussed in Chapter 1. The difference is that an NPCS has been evaluated and the potential hazards either are not present or have been eliminated through engineering controls. Knowledge of this differentiation is intended for workers who will enter the space and perform their work. It is not intended for rescuers. Rescuers responding to a confined space accident must initially treat all confined spaces as permit-required spaces. Only after careful evaluation of the space should rescuers consider a confined space a non-permit required space.

NOTE
To help ensure that the hazards (both atmospheric and physical) are reviewed prior to entry, certain requirements must be met for a basic confined space entry permit.

Prior to issuing a confined space entry permit, numerous safety concerns must be reviewed. This review includes knowledge of the following:

- The location of the space
- The reason for the entry
- Atmospheric monitoring (inside and in proximity to the outside the space)
- Personnel Protective Equipment (PPE) requirements
- Lockout/tagout requirements
- Other potential hazards that the space may contain

Some of these hazards may not exist within or near the space, some can be eliminated, and still others will be present and must either be controlled or the entrant protected from them. The only way to effectively manage the numerous review items is to develop and use a standardized checklist at the confined space, as shown in Figure 2-7. This checklist is the basis for the confined space entry permit. To help ensure that the hazards (both atmospheric and physical) are reviewed prior to an entry, certain requirements must be met for entry and recorded on the permit. These requirements can range from the simple to the technical, but the problems are no less significant for the simple items. The correct identification of the confined space may seem fairly simple but with an incorrect identification, everything else is incorrect. If rescuers take simple items for granted, they may miss the simple problems. Written documentation in the form of a confined space entry permit is what you, as a rescuer, should ask for and review as soon as you get to a confined space entry emergency. If the permit is present, you can begin verifying the information on it. A missing permit is the worst case scenario because there is no confined space entry program, and you will have to determine all of the information you need to make a rescue.

Model confined space entry permits can be found in CFR 29—Standard 1910.146, the OSHA Permit-Required Confined Spaces standard and ANSI Standard Z117.1-2003, *Safety Requirements for Confined Space*. Other examples of confined space entry permits can be found at various locations on the Internet, but you should be certain that these examples meet the minimum requirements of the OSHA standard. NFPA 1006, *Standard for Rescue Technician Professional Qualifications,* 2003 edition contains sample confined space entry permits for use by the rescue team.

A confined space entry permit that meets the intent of the OSHA standard should address the following information:

- Date and time the permit was issued and the date and time the permit expires. No permit should be issued for an indefinite time period. Conditions change over time, and the changes can originate inside the space or outside the space. An outdated permit invites disaster. It gives workers a false sense of security, which results in them working in an unsafe environment. Some permits are issued for as long as a year, but such a long timeframe must be based on an evaluation of that specific space, the potential hazards, and the very low potential for hazards to be introduced into the space.
- Accurate identification of both the job site/space and the supervisor. Employees must know that they are entering the correct space and that they have properly identified the hazards of the space they are entering. The supervisor must also be identified so that workers can get answers to any questions they may have about the space, including the signs and symptoms of exposure and the potential hazards.
- The equipment to be worked on and the type of work to be performed. It is important to make the space safe from unexpected and deadly hazards, such as accidental start-ups of equipment, the accidental opening of a valve, applying a combustible coating inside the space, and so on. By knowing what type of work is to be performed and where it is to be done, as well as work that was not planned for and work the space was not evaluated for, the workers can protect themselves and others from injury or death.
- Correct identification of the attendant and the entrants. Entrants, attendants, and other workers at the job site must know who belongs there and what is expected of them.
- Atmospheric checks before and during entry. Atmospheric hazards are the leading cause of

```
                                    ENTRY PERMIT
        PERMIT VALID FOR 8 HOURS ONLY, ALL COPIES OF PERMIT WILL REMAIN AT JOB SITE UNTIL JOB IS COMPLETED
        DATE  -  - SITE LOCATION and DESCRIPTION_____
        PURPOSE OF ENTRY_____
             SUPERVISOR(S) in charge of crews            Type of Crew            Phone #
        _____
        _____
        COMMUNICATION PROCEDURES _____
        RESCUE PROCEDURES (PHONE NUMBERS AT BOTTOM)_____
        _____
        *BOLD DENOTES MINIMUM REQUIREMENTS TO BE COMPLETED AND REVIEWED PRIOR TO ENTRY*
        REQUIREMENTS COMPLETED              DATE             TIME
        Lock Out/De-energize/Try-out
        Lines(s) Broken-Capped-Blanked
        Purge-Flush and Vent
        Ventilation
        Secure Area (Post and Flag)
        Breathing Apparatus
        Resuscitator - Inhalator
        Standby Safety Personnel
        Full Body Harness w/"D" ring
        Emergency Escape Retrieval Equip
        Lifelines
        Fire Extinguishers
        Lighting (Explosive Proof)
        Protective Clothing
        Respirator(s) (Air Purifying)
        Burning and Welding Permit
        Note: Items that do not apply enter N/A in the blank.
                    **RECORD CONTINUOUS MONITORING RESULTS EVERY 2 HOURS
        CONTINUOUS MONITORING**          Permissible
        TEST(S) TO BE TAKEN              Entry Level
        PERCENT OF OXYGEN                19.5% TO 23.5%
        LOWER FLAMMABLE LIMIT            Under 10%
        CARBON MONOXIDE                  +35 PPM
        Aromatic Hydrocarbon             +1 PPM*5PPM
        Hydrogen Cyanide                 (Skin)*4PPM
        Hydrogen Sulfide                 +10PPM*15PPM
        Sulfur Dioxide                   +2PPM*5PPM
        Ammonia                          *35PPM
        *Short-term exposure limit: Employee can work in the area up to 15 minutes
        +8 hr. Time Weighted Avg. Employee can work in area 8 hrs (longer with appropriate respiratory protection).
        REMARKS:
        _____
        GAS TESTER NAME &    INSTRUMENT(S) USED    MODEL &/OR TYPE    SERIAL &/OR UNIT #
        CHECK
        _____
                    SAFETY STANDBY PERSON IS REQUIRED FOR ALL CONFINED SPACE WORK
        SAFETY        CHECK #     CONFINED        CHECK #     CONFINED        CHECK #
        STANDBY                   SPACE                       SPACE
        PERSON(S)                 ENTRANT(S)                  ENTRANT(S)
        _____

        SUPERVISOR AUTHORIZING     ALL CONDITIONS SATISFIED _____
                                   DEPARTMENT/PHONE _____
        AMBULANCE 2800    FIRE 2900    SAFETY 4901    GAS COORDINATOR 4529/5387
```

FIGURE 2-7 This is a sample confined space entry permit.

death in confined spaces and must be among the most critical items evaluated prior to and during the confined space entry. How could one person begin monitoring before entry, provide continuous monitoring during the entry, and remember all of the different readings without some type of checklist? The entry permit requires that readings be taken for oxygen levels, flammable gases (also called explosive gases), and toxic gases based on the known or suspected hazards in the space. It is important that the person checks the operation of the monitoring equipment before use and starts taking the readings outside of the space. Equally critical is to take the readings within the space either simultaneously or in the following order: oxygen, flammable/explosive limit, and toxicity. The tester must also sign or initial the permit to confirm that the readings have been taken and recorded.

> **NOTE**
> Atmospheric hazards present the most common risk to people in a confined space.

- Atmospheric hazards are the principal cause of death in a confined space, and engulfment is the second leading cause. Preventing engulfment requires eliminating and securing sources of potential physical or chemical hazards by isolating them from the space and/or controlling flowable materials in the space. This procedure is called **lockout/tagout** and will be discussed in greater detail in Chapter 4. Included in lockout/tagout procedures are items such as eliminating the material from the space, shutting down pumps, capping product lines that enter the confined space (also called **blinding**), and disconnecting or blocking the lines.
- Controlling atmospheric hazards and the need for ventilation. It may be necessary to provide mechanical ventilation for the space. If the hazards can be eliminated, it may be possible to reclassify the confined space as a non-permit required space and use natural ventilation. When the only means of controlling an atmospheric hazard is by using forced ventilation, the space cannot be reclassified, and mechanical ventilation must be in place.
- When either isolation or ventilation is used, a thorough check must be done to ensure that the ventilation and isolation are effective. After the confined space has been properly isolated and/or vented, someone must take meter readings for oxygen, combustible gases, and toxic gases to determine the isolation and/or venting have been effective. The acceptable levels of each atmospheric hazard should be identified on the permit. For oxygen, a level between 19.5 and 23.5 percent in air is required. If combustible gases are present, the level must be less than 10 percent of the lower flammable limit for that product. Toxic gases must not be at present a level greater than the **permissible exposure limit (PEL)** of the toxic material in air.
- Communication between the attendant and entrant(s) is critical to confined space safety. Communication procedures must be outlined on the permit so that the attendant can check with the entrants. This becomes especially important if an entrant will have to perform work out of site of the attendant. The attendant must not enter the confined space for any reason unless another trained person takes the role of attendant. By OSHA definition, entry is considered to have occurred *as soon as any part of the entrant's body breaks the plane of an opening into the space.* This means, for example, that if the attendant leans into the opening or puts her face into the opening to maintain contact with an entrant, she has made entry and is no longer considered an attendant, but rather an entrant. Effective communication procedures protect both the attendant and entrants.
- Rescue procedures must be outlined on the entry permit. This is the beginning of your involvement. When your group or agency has been designated as the rescue team, you have the right to inspect the confined space so that you can preplan a rescue. If your agency or group is not the designated rescue team, you must understand what your role at that emergency is. Is it a support role? Or have you been called because there is no rescue plan? Knowing in advance about the types of on-site confined spaces and the hazards they present will save time and lives—maybe yours. Cooperation and preplanning are keys to an efficient operation. NFPA 1006 contains a basic sample confined space rescue response agreement to provide rescue services to individual employers. Any specific concerns that need to be addressed can be added into this agreement.

> **NOTE**
> If your group or agency is the designated rescue team, then you have the right to inspect the confined space so that you can preplan a rescue.

- Identification of all entry, standby, and backup persons on the permit. Part of the confined space program should be to keep unqualified people out of confined spaces. Not only must the entry, standby, and backup persons be identified, but they must also have successfully completed all required training, and it must be current. Consider this example showing why a permit is so valuable for the entry. Imagine if you were the entry supervisor and had to remember all of the items without a checklist. Do you think that you could accomplish this task accurately enough to protect another person's life?

Not only does a permit-required entry call for different types of equipment, but the equipment must also be at the worksite. A checklist prompts employees and supervisors to avoid taking shortcuts. The equipment should also be functioning properly. What good is a combustible gas detector if it has dead batteries? What about a defective self-contained breathing apparatus (SCBA)? Not having functional protective equipment at a confined space entry worksite can actually cause an accident when all other precautions have been taken. Among the equipment that needs to be considered for use at a confined space entry are the following:

- Direct-reading gas monitors, as shown in Figure 2-8
- Safety harnesses and lifelines for entry and standby persons, as shown in Figure 2-9 hoisting equipment
- Communications Equipment

FIGURE 2-9 Harnesses allow rescuers to be lowered into a confined space and retrieved should they become incapacitated. Ladder belts, like the one shown in the right of the picture, should not be used to raise or lower a person.

- SCBAs for entry and standby persons
- Protective clothing, as shown in Figure 2-10

FIGURE 2-10 These are spare air cylinders specifically set up for a supplied air respirator. Having rescue and safety equipment that is designed for use at a confined space rescue improves rescuer safety and enhances operations.

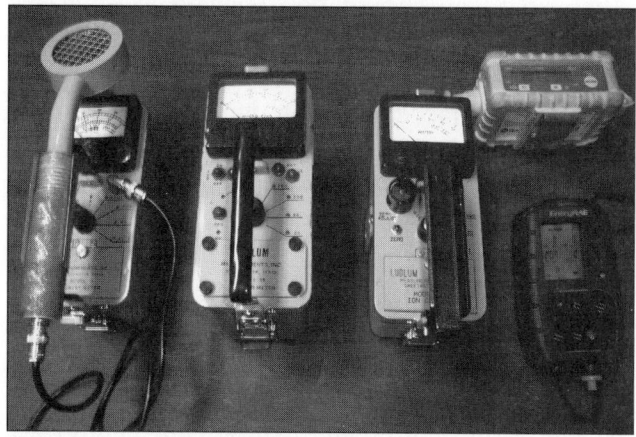

FIGURE 2-8 Shown here are several different types of direct reading monitoring devices.

- Electric equipment specified as **Class I, Division I, Group D**
- Non-sparking tools

> **NOTE**
> If the confined space atmosphere is monitored, readings are below 10 percent of the LEL, and ventilation is maintained, then the non-sparking tools may not be needed. This decision should be made by the entry supervisor.

Though much of the equipment listed here is intended to prevent an accident based on potential atmospheric hazards in the confined space, it does not eliminate the need for periodic atmospheric testing or monitoring. You must maintain a safe atmosphere during a confined space entry. During a rescue, you will want to monitor the space to determine what hazards might be present, continue monitoring to make certain the conditions are not getting worse and, if you are ventilating the space, make sure your ventilation is effective. The type of monitoring you employ and the frequency is going to vary from space to space and will be based on the anticipated hazards. Non-sparking tools and appropriate, listed electrical equipment are examples of ways to decrease the atmospheric hazards by controlling potential ignition sources. But all of the hazards are not atmospheric. There are also physical hazards in and around a confined space. You need to look in and around the confined space so that you can recognize physical hazards such as electrocution from damaged extension cords and electrical tools, requirements for chemical protective clothing, stable hoisting equipment that is designed to be used to hoist people into or out of the space, and hoisting equipment to raise or lower tools and materials into the confined space. One word of caution about hoisting equipment: Use only manually powered hoists for raising and lowering people in and out of a confined space. Without proper safety devices (that is, a clutch), power-driven hoisting equipment can operate at too fast a speed and create enough force to injure a person who gets caught on something within the space while being hoisted. Manually powered equipment may be preferable because the equipment can be operated at a slow enough speed to avoid producing forces that can easily injure the person being hoisted.

> **NOTE**
> Use manually powered hoists for raising and lowering people in and out of a confined space. Powered equipment may not be used unless it has been designed for raising or lowering people and is equipped with the proper safety devices, that is, a clutch or similar mechanism to stop motion once a predetermined level of resistance is reached.

Someone has to sign the permit to show they have reviewed it and approve entry under the conditions listed on the permit. This signatory is the entry supervisor or other qualified person in charge of the entry (for example, rescue entry supervisor, incident commander, rescue team leader, etc.). OSHA regulations do not require all entrants to sign the permit; however, all entrants should review the permit to confirm that it has been filled out and that conditions are acceptable to enter. Some employers use internal confined space entry procedures that exceed OSHA requirements. These internal permits may contain a space at the end of the permit for the employees who will be entrants and/or attendants to sign. These signatures are for company purposes only. Whether or not employees sign the permit, all employees exposed to the hazards of a confined space should receive written instructions and be made aware of the required safety procedures. They must understand the hazards and precautions so that they can protect themselves and others. No one should be expected to enter the confined space if any of the questions on the permit are answered in a way that indicates the presence of an unacceptable or uncontrolled hazard. All questions on the permit should be answered yes, no, or not applicable. Keep the questions and answers on the permit simple so that the answers are not confusing. Questions with a no answer should indicate that a hazard or condition exists that has not been properly addressed, and entry conditions are not acceptable. The entry permit is not valid unless all appropriate items required by the permit are completed. Employees should not treat this section of the permit as an administrative formality. Identifying potentially hazardous items in the space is intended to remind employees of their right to work in a safe manner. Other signatures at the end of the permit identify the person who prepared the permit (the entry supervisor) and the unit supervisor in whose area the confined space is found.

> **NOTE**
> Depending on internal requirements within certain companies, there may be a space at the end of the permit for the employees who will be entrants and/or attendants to sign. These signatures re intended for internal use only.

The employer must keep a copy of the permit at the job site, as shown in Figure 2-11. This written permit is a valuable source of information to emergency responders. The permit must be on-site as long as entry operations are ongoing. Once the original work is completed, a copy of the permit must be kept at the facility for a year after the entry. Knowing this is valuable to you as a rescuer in the event you are called to a confined space accident where the work was completed but someone re-entered the space without a valid permit. Reviewing the original permit may help you in evaluating the situation.

ANNUAL CONFINED SPACE ENTRY PERMIT

Location: Primary Pump Station - Dry Well   Date Posted: June 19, 1999

Permit Expires: June 18, 2000

Description of Work Normally Performed: Routine Operator check, maintenance and cleaning.

Persons Authorized to Enter (Positions): Operations and Maintenance Departments personnel and authorized visitors with escorts.

Atmospheric Testing (Perform Prior to Posting)

Time: 0846   Oxygen Level: 20.9 %

Make of Unit: CGM   Toxic Level: 0 PPM

Serial No: 9161   Combustible Level: 0 %

Atmospheric Testing Performed by: George Browne/ *George Browne*
Print/Signature

Above Results Confirmed by: Gus S. Crist/ *Gus S. Crist*
Print/Signature

Any special instruction for this entry?: Yes: X   No: ___
(i.e., gas detector, ventilation, etc.)

Note special instructions here: When ventilation is on, no gas detector is required. When ventilation is not operational, a gas detector is required for entry.

Note: This Annual Permit will be revoked whenever any conditions in the confined space have become more hazardous than contemplated.

Supervisor Authorizing this Annual Permit: Gregory Hansson/ *Gregory Hansson*
Print/Signature
Gregory Hansson, Director, Eastern Division

h:\ops\confspps.xls   7/99

FIGURE 2-11 A copy of the confined space entry permit must be kept at the job site. This particular permit is a long-term permit (issued annually) and spells out what work is to be done and who is to do it.

The permit can specify PPE and rescue equipment. Spare air cylinders are specifically set up for a supplied air respirator. Having rescue and safety equipment that is designed for use at a confined space rescue improves rescuer safety and enhances operations.

CONFINED SPACE PROGRAM AND RESCUERS

As a rescuer reviewing a confined space entry permit, you may think that it is too complicated and not of value you during an emergency. It is not complicated, just new to many rescuers. As part of a rescue team, you must understand the roles of the attendant, entrant, and supervisor. You must be able to identify the key roles that rescuers must fill to provide for the safety of the rescue team. NFPA 1006 *Standard for Rescue Technician Professional Qualifications* 2008 Edition makes no differentiation between the definitions of authorized entrant and rescue entrant, or attendant and rescue attendant. You might not think that the rescuers who will enter the confined space are entrants, but they are. An equivalent level of knowledge about the confined space and the hazards it contains is required by the rescue entrant. Your team might call the rescue attendant the safety officer or use some other name instead of attendant, but does the safety officer perform functions that are different from the attendant? Isn't that person assigned to monitor the confined space and the people who enter during a rescue? The role of the entry supervisor might be performed by the person who is identified as the rescue team leader or, during simple rescue operations, the incident commander. At other times, the entry supervisor could also be the safety officer, and then the rescue attendant's role might change to simply monitoring the rescuers in the space. The point being made here is that the title is not as important as identifying the needed tasks and assigning people to perform them. Comprehending the requirements of a confined space entry program and adapting those requirements to rescue purposes protects your team members, as well as you. Keep it simple by identifying what steps you must take to guarantee your safety, protect the victim, and determine what equipment you will need to perform your rescue operations. Look at the requirements of the various confined space entry standards and confined space rescue standards (OSHA, ANSI, NFPA). Use those standards to define your needs and then plan on meeting those requirements or identify the limits of what your team can do.

PERMIT-REQUIRED VERSUS NON-PERMIT CONFINED SPACES

Chapter 1 discussed permit-required and non-permit confined spaces. If you have a permit-required confined space, you must have an entry permit present. But what happens with a non-permit confined space? If the non-permit space is expected to be free of hazards, the employer does not need a confined space permit. As a rescuer, you will not find a confined-space permit at a non-permit space worksite, as shown in Figure 2-12. Don't take for granted that the lack of a permit means the space does not contain any hazards. The space may have originally been a non-permit required space, but you should not ignore the precautions laid out in the OSHA confined space entry program. Non-permit spaces may have had an unidentified or unexpected hazard introduced because of the work being performed in it. Simple work such as painting the inside of a space can lead to a lowered oxygen level or the presence of toxic vapors. Your

FIGURE 2-12 Not all spaces are permit-required confined spaces. This space does not require a confined-space entry permit because it does not meet the definition of a permit-required confined space. But if an emergency occurred, how would you handle this space?

training must create a sense of awareness and caution that leads you to initially treat the space as permit-required, using basic precautions (for example, air monitoring, ventilation, lockout/tagout), assessing the physical hazards, and determining that the area is free of hazards. If you can't be certain that the space does not require a permit, treat it as a permit-required space.

> **NOTE**
> Just because the space was originally a non-permit required space does not mean that you should ignore the precautions laid out in the confined space entry program.

LESSONS LEARNED REVISITED

The excavation is 15 feet square and the depth is 27 feet, and the ground marshy with a high water table. The pit was shored with vertical steel sheeting that extended above the top of the excavation. An unsecured ladder leads to an intermediate ledge (called a wale), and a second unsecured ladder leads from the wale to the bottom of the pit. The excavation site includes pumps that were used to pump water from the pit, but they were not running at the time of the accident. Looking into the hole, you see a construction worker lying at the base of the excavation and another worker lying nearby. Water and sand are "boiling" from a partially driven piling at the bottom of the hole, and the smell of rotten eggs permeates the area.

Discussion Questions
1. If you were the confined space entry supervisor, what precautions would you require prior to allowing workers to enter the space?
2. Would you require the use of a harness and retrieval equipment during entry?
3. Hydrogen sulfide is recognizable by its rotten egg smell, and it is both toxic and flammable. What other characteristics of this incident site would lead you to anticipate the presence of hydrogen sulfide?

SUMMARY

- The purpose of a confined-space entry program is prevention of injuries and deaths.
- If you have been called to a confined-space accident, something has gone wrong at the site and you must determine what has happened.
- Emergency workers can use the site's confined space program and confined space permit to assess what has happened. A properly completed permit will assist you in determining the potential hazards of the confined space. Even though the problem may not be identified on the permit, it still provides a starting place. There should be an attendant, supervisor, and possibly an entrant at the scene. You should interview these people to find out what has happened.
- When there is no confined-space program for the workers and an emergency has occurred, you will be at a disadvantage and will have to develop the information on your own.

- Understanding the basis of a confined space entry program should make you aware of the need to structure your rescue operations in a similar fashion.
  - Chapter 9 details how to implement a confined space rescue program based on a real program.
  - The roles of attendant, entrant, and supervisor change only slightly for rescue.
  - For rescue operations, all confined space emergencies should be considered to involve permit-required spaces.
  - Only after you are done monitoring and investigating the conditions within the space and clearly identifying it as safe enough to downgrade to a non-permit level should you change the classification of the space.

REVIEW QUESTIONS

1. The role of the attendant is to stay outside of the confined space and monitor conditions both inside and outside the space. Explain the value of monitoring such conditions.
2. The attendant is not to enter the confined space for any reason. Why is this requirement so critical to the safety of people within the space?
3. Briefly explain how the roles of attendant, authorized entrant, and entry supervisor would be paralleled by the members of a rescue team.
4. Identify four pieces of information that must be contained on a confined space entry permit and how the information is valuable to a rescue team.
5. Explain the difference between a permit-required confined space and a non-permit confined space. How would this difference affect your rescue operations?
6. What is the main reason for an entry permit?
    a. To identify the number of personnel working in the confined space.
    b. It is a written document that outlines what is going on in the space.
    c. It is a sign-off sheet that everything is safe before entering.
    d. All of the above.
7. The attendant is authorized to bring tools into the confined space when performing the attendant duties. True or False?
8. To isolate a physical or chemical hazard, you need to _____ the area in question.
9. You should consider using an electrical or hydraulic hoist to raise or lower people in a confined space. True or False?
10. Who is in charge of a confined-space entry?
    a. entrant or entrants
    b. attendant
    c. confined space supervisor
    d. all of the above

KEY TERMS

entry permit 15
attendant 16
Standby Assistance 17
attendant-based rescue 17
authorized entrants 18
confined space supervisor 18

rescue attendant 18
rescue entrant 18
lockout/tagout 22
blinding 22
Permissible Exposure Limit (PEL) 22
Class I, Division I, Group D 24

ACTIVITIES

1. Arrange a tour of a local facility with a confined space entry program.
2. Obtain MSDSs or other hazard information for various materials (for example, methane, hydrogen sulfide, epoxy paints) that may be present or brought into a confined space. Using those information sources, evaluate the materials, their characteristics, and how they can be hazardous within a confined space.

ADDITIONAL RESOURCES

There are various groups with experience in confined space entry program. For more information on the topics discussed within this chapter, you can go to the following:

OSHA representatives, http://www.osha.gov/ NIOSH, http://www.cdc.gov/niosh/

The State Department of Labor National and local chapters of the following organizations:

- American Society of Safety Engineers, http://www.asse.org/
- American Institute of Industrial Hygienists, http://www.aiha.org/Content

State or local agricultural agents or agencies Local facilities, including:

- Water treatment plants
- Local facilities where you are part of the designated rescue team
- Local colleges and universities with health and safety programs

3 Air Monitoring

LESSONS LEARNED: TWO FOREMEN DIE IN RAIL CAR ACCIDENT

Two foremen employed by a company that repaired railroad cars died from asphyxiation after entering a tank car. The tank car had been involved in a derailment, was damaged, and then was sent to this facility for repairs. At the time of the derailment, the car had contained soybean oil, and the unfilled space (headspace) was filled with nitrogen to keep the oil from spoiling during shipment. The rail car had been off-loaded prior to being sent to the repair facility, but it was not marked as a permit-required confined space at the time of the incident. One of the victims, an experienced foreman, used a multi-gas meter to test the atmosphere inside the cargo space of the railcar and told the attendant that the atmosphere was fine. He then entered the space and collapsed a short time later.

The attendant called for help, and the second foreman arrived. The attendant told the second foreman what the first foreman said about the air quality, and the second foreman entered thinking it was a medical emergency. Following the second foreman into the rail car, the attendant was on the ladder when he saw the second foreman collapse. At this point, the attendant exited the space and waited for rescue crews to arrive.

Critical Thinking Questions

1. Based on the characteristics of the rail car in the preceding incident, would you identify the tank as a confined space?
2. Would you consider the rail car a permit-required confined space?
3. If you consider the tank to be a permit-required confined space, what OSHA requirements should be in place prior to entry?

LEARNING OBJECTIVES

NFPA 1006, *2008 EDITION*
By the end of this chapter, you should be able to:

- Describe the capabilities and limitations of the following monitoring equipment (7.1.1):
  - Combustible gas indicator
  - Oxygen meter
  - Specific gas monitoring equipment
  - Gas-specific meter
  - Colorimetric tubes
  - pH devices
- Define the following terms:
  - Parts per million
  - Percentage of the lower flammable limit
  - Action limit
  - pH scale
  - Confined space configuration
  - Calibration

INTRODUCTION

You are standing on the bank of a river swollen with flood waters. In the middle of these raging waters, you see a person trapped in the river and they cry for help. As you look at the muddy brown water, it charges past with a force that is pushing along debris as large as a small car. What hazards immediately come to your attention? Would the hazards prevent you from entering the river to rescue this trapped person? Let's change the situation around and, instead, imagine that you are at a chemical plant. You and your rescue team have been called there for a confined space accident. As you stand at the opening of a process vessel, two workers lie unconscious at the bottom, and their coworkers are anxiously asking you to rescue their friends. The victims are not trapped by any of the equipment within the space, and the scene inside the vessel is eerily quiet. What hazards immediately come to your attention? Can you identify the dangers that are present? Unless there is a visible vapor cloud within the confined space, you will not see an atmospheric hazard. You usually cannot visibly detect atmospheric hazards, and you cannot accurately determine the degree of danger present. How do you establish the degree of atmospheric hazard present? This is where monitoring equipment plays an important role. Using the right monitoring equipment allows you to qualify or quantify the existing hazards, to determine the level of danger the hazards present to both the rescuers and the victims, and to begin a risk to benefit analysis for the anticipated rescue operation. Before you begin to monitor for atmospheric hazards, you must be familiar with the types of atmospheric hazards that may present themselves in a confined space, the types of monitoring equipment available to detect and/or measure those hazards, the limitations and capabilities of that equipment, and the requirements for calibration and maintenance.

NOTE
You qualify something by defining a feature or characteristic that meets a general definition but may not be measurable. When you quantify something, you measure it to determine its magnitude, sum, or level against a specific scale.

There are many types of atmospheric monitoring equipment available. In this book we will discuss only direct reading, real-time instruments for use in confined spaces. As the name implies, **direct reading instruments** provide a reading either from a digital readout, a gauge of some type, a measurable change of color of a sample tube, or some other decipherable indicating method. What is important is that the results are measurable, and you can interpolate the data you are being presented with. The measurement might be presented in parts per million, as a percentage, as pH (a logarithmic scale), or via some other scientifically recognized measure. Some monitoring devices can detect only the presence of a particular gas and cannot provide a direct measurement. The value of this type of instrument is limited because it qualifies the atmosphere as containing a particular gas (for example, CO, natural gas, oxygen) but does not quantify it by letting you know how much of that gas is present. Such an instrument is useful as a warning device to alert people to the need to evacuate (a CO detector is a good example), but it is of little value to rescuers who need to know how much CO is present so that they can develop an action plan and select the proper equipment. In addition to providing a direct reading, your instruments should give real-time or immediate reading. Therefore, you will experience no time delay because a sample had to be sent for analysis at a laboratory. Emergency responders need real-time and direct reading instruments to provide immediate, on-scene measurements, as shown in Figure 3-1.

> **NOTE**
> For emergency responders, real-time and direct reading instruments provide immediate, on-scene measurements.

It would not be possible for this text to provide a complete guide to the use of monitoring instruments. Only by consulting the manufacturers' instructions and training with your monitoring instruments well in advance of an emergency can you hope to become proficient in using the instruments and interpreting the information you are getting from the equipment.

FIGURE 3-1 The instrument shown in the center of this picture is a combustible gas detector combined with a multiple gas detector. Surrounding the gas detector is calibration equipment for use with the detector.

The Lessons Learned section of this chapter provides an example of what can happen if you fail to monitor properly or fail to maintain your equipment. You must know not only how to use your instruments, but also the limitations of the equipment and the maintenance requirements. All monitoring instruments require periodic calibration, replacement of sensors and batteries, and testing of the operation of the equipment. The manufacturer is the best source to provide you with the proper operation, inspection, testing, and maintenance of equipment.

> **NOTE**
> By using the right monitoring equipment you can begin to characterize the type of hazard present and the degree of danger it presents to both the rescuers and the victims.

COMBUSTIBLE GASES

To help detect and measure the presence of combustible or flammable vapors, use a **combustible gas indicator (CGI)**, as shown in Figure 3-2. The CGI may also be referred to as an explosimeter, and it measures what percentage of the lower flammable limit is present in the atmosphere. Understanding how a combustible gas

Air Monitoring 33

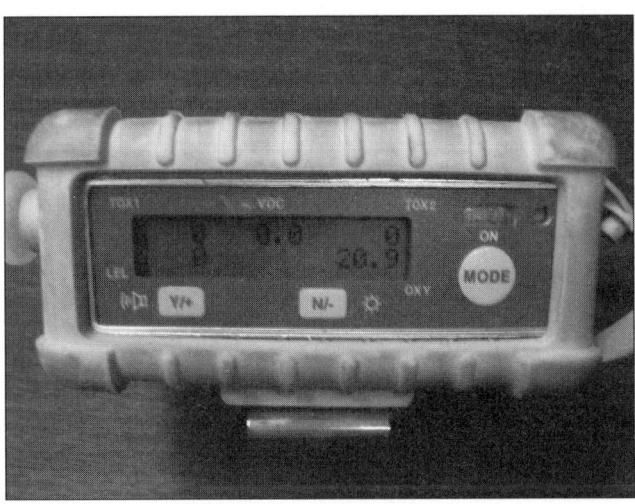

FIGURE 3-2 Combustible gas detectors indicate how close the gas concentration is to the lower flammable limit. A reading of 100 percent indicates that the concentration is at least at the lower flammable limit. This meter is showing a 0 percent LEL reading.

indicator operates requires that you first understand that it is an electrical device that uses a balanced electric circuit known as a Wheatstone Bridge. The meter reading is essentially a measure of the change in resistance within the Wheatstone Bridge that is translated into a percentage of the LFL. One side of the Wheatstone Bridge circuit is in the control chamber. The other side of the circuit is in the test chamber. When an atmospheric sample is drawn into the meter, it comes in contact with a heated wire in the test chamber. As the sample is heated in the test chamber of the meter, as the sample reacts with the heated wire, it heats that side of the Wheatstone Bridge, which causes the resistance in the electric circuit to change. This change causes an electrical imbalance between the legs of the Wheatstone Bridge, which can be read by the meter. The more out of balance the circuit, the greater the meter reading and consequently the higher the percentage of LFL that is considered to be present.

Before you use a CGI, there are some important questions to consider about the instrument. To begin, what does the instrument measure? Combustible gas indicators do *not* tell you the percentage of combustible gas in air. They do not tell you how close the gas concentration is to the LFL for the gas you are measuring. Chapter 1 presented a discussion of the lower flammable limit, the upper flammable limit, and the flammable range. Below the LFL, the gas mixture in air is too lean to burn because it lacks adequate fuel. Between the LFL and the UFL (the upper flammable range), the fuel and air mixture will support combustion and, with an ignition source present, the mixture will ignite and burn. Above the UFL, the mixture simply contains too much fuel. Remember that not all gases have the same flammable range and that variation must be recognized and understood.

You also need to understand that the combustible gas indicator gives a reading in percentage, but a percentage of what? Figure 3-3 shows a reading of percentage of the lower flammable limit and is an accurate reading. A reading, for example, of 25 percent indicates that the meter detected a level of combustible gas that was one-fourth of the way to the lower combustible limit. A meter reading of 100 percent indicates that the meter had reached 100 percent of the lower combustible limit. Importantly, a meter reading of 100 percent does not mean that the concentration of combustible gas in the air is at a 100 percent. If you do not recognize this fact, you could easily believe that the meter reading reflects the concentration of gas in air. The assumption that there is a 100 percent concentration of gas in air is incorrect, and with a 100 percent reading you might even believe that the gas was above its UFL and therefore too rich to burn. In fact, the gas is at least at the LFL, may well be within the flammable range and only needs an appropriate ignition source to ignite it. Failing to interpret

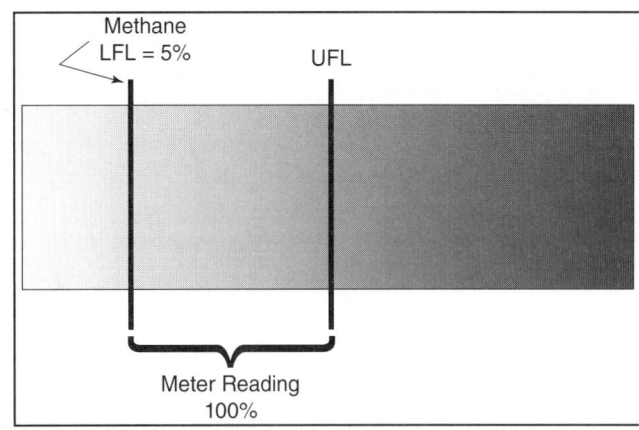

FIGURE 3-3 Combustible gas detectors indicate how close the gas concentration is to the lower flammable limit of the gas with which the meter is calibrated.

the information correctly can leave you with a mistaken impression of the hazard.

As stated previously, not all gases have the same flammable range or the same lower flammable limit. You must know what gas was used to calibrate the meter as well as the flammable range and the lower flammable limit of that calibration gas. Methane is often used for calibration, and it has a lower flammable limit of 5 percent methane in air. Pentane is also used to calibrate combustible gas detectors, and it has a lower flammable limit of 1.5 percent pentane vapor in air. As a result of the use of pentane for calibration, a pentane calibrated meter would give a greater percentage of the LFL reading in the same flammable gas/air atmosphere than a meter calibrated using methane.

So far we have discussed the CGI, how it works, what information it is giving you, and the gas with which it is calibrated. Now we need turn our attention to the properties of the combustible gas you are trying to detect. If you are using a CGI calibrated on methane (LFL 5 percent) and the gas you are trying to measure has an LFL of 2 percent in air, will the meter be able to effectively tell you when you have reached the 100 percent mark on the CGI? The simple answer is no. The CGI is meant to measure the percentage of LFL for the gas it is calibrated with, as shown in Figure 3-4. When the LFL of the calibration gas is 5 percent, and you have a gas present with an LFL of 2 percent, the LFL of the calibration gas is 2.5 times the LFL of the gas you are measuring. The gas on hand will be well into the flammable range before the CGI reads 100 percent. In addition, you may not be able to identify the combustible gas that is present and will not know it characteristics. To compensate for this situation, the CGI should go into alarm once it reaches 10 percent of the LFL of the calibration gas (0.5 percent methane in air). This is considered an **action limit** and should be part of your emergency response procedure—whenever you reach a 10 percent LFL reading on the CGI, all activities will stop in the area containing the flammable gas or vapors and anyone in that hazardous area will leave that area. This provides a margin of safety that will generally keep you out of a 100 percent LFL (or higher) atmosphere. You must then eliminate or control the hazard (for example, by using ventilation) and bring the CGI reading below 10 percent. This process serves two purposes: (1) you are building in a safety factor to make up for the limitations of your equipment and (2) staying at a level below 10 percent of the LFL alerts quantifies the effectiveness of your hazard control efforts.

> **NOTE**
>
> The CGI is meant to detect the LFL of the gas it is calibrated with, not the percentage of gas in air.

Also be aware that certain meters can identify a 100 percent LFL reading and then immediately return to a reading of 0 percent LFL. This is more common with older meters and can occur when the meter is used in an atmosphere above the UFL of the calibration gas. What happens is that the meter detects an atmosphere of 100 percent LFL and when the gas drawn into the meter goes above the UFL, it is too rich to combust within the meter and heat the wire. Because the meter is an electrical device and no heating of the wire occurs, the Wheatstone Bridge circuit is back in balance and the meter reads 0 percent.

As you use a combustible gas indicator, or any metering device, you must take into account the differences in vapor density of different gases. This means you must sample the area at different

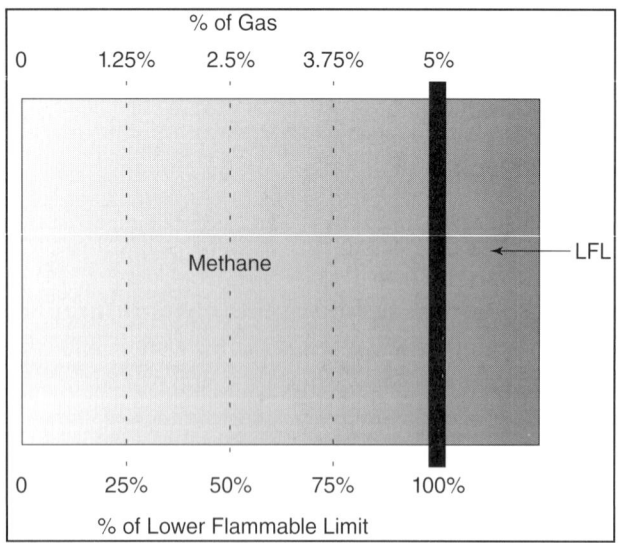

FIGURE 3-4 For a CGI calibrated on methane, a meter reading of 50 percent would indicate a gas concentration equal to 2.5 percent methane in air.

elevations. Gases that are heavier than air settle and can be expected to collect at lower levels. Lighter than air gases rise and can be expected to collect at upper levels, as shown in Figure 3-5. Metering the space at only one level will cause you to miss gases that are more concentrated above or below the meter level. Avoid this problem by sampling at different levels so that you meter the space at the top and bottom as well as the intermediate levels.

> **NOTE**
> You must meter the space at the top and bottom as well as the intermediate levels.

Attempting to interpret the information you obtain while using a CGI can cause confusion. To help you understand and use the information you collect, consider the following scenarios:

1. Your combustible gas indicator is calibrated with methane (LFL of 5 percent gas in air). Natural gas is known to be in the confined space, and you have a meter reading of 50 percent. Is this 50 percent methane in air or 50 percent of the way to the lower flammable limit?

This is a concentration of 50 percent of the LFL, and the meter was calibrated using methane (natural gas). It is also 50 percent of the LFL of the gas in the space. If you reach a level of natural gas that is 50 percent gas in air, your meter reading should be 100 percent because it is above the LFL of methane.

2. The CGI you are using in this scenario is calibrated using pentane, which has an LFL of 1.5 percent in air. The meter sounds an alarm at 10 percent LFL. As you are using the meter, the meter alarm sounds again, and the reading is 15 percent LFL. Is it safe to continue to work in this area, or should you leave and try to remove the hazard?

In this scenario, 15 percent of the LFL of pentane is 0.225 percent gas in air. Whether or not you have identified the type of gas you are detecting with the CGI, you still may not know the LFL of the gas. As mentioned previously, when using a CGI, you should have an action limit of 10 percent and stick to it. That means whenever you get a 10 percent LFL reading on the meter, anticipate that the atmosphere is hazardous and take steps to control or remove the hazard. Even though

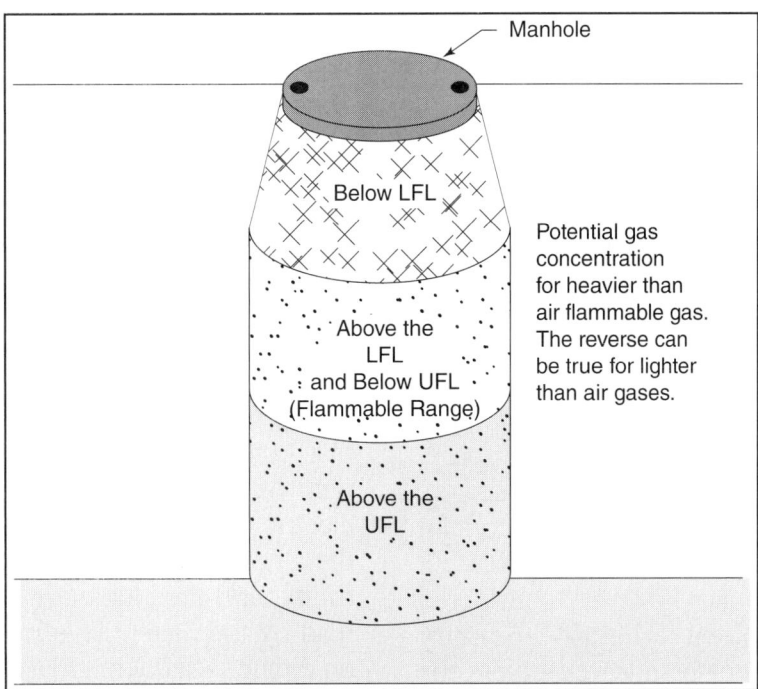

FIGURE 3-5 Depending on the vapor density of the gas you are attempting to monitor, the gas concentration may vary within the space. A gas that is heavier than air will tend to be more concentrated at the bottom of the space. The reverse would be true for a gas that is lighter than air.

you are using a meter calibrated on a gas such as pentane and the low LFL, you must accept that a 10 percent meter reading is a strategic factor. Strategic factors limit or affect your options and actions.

Combustible gas indicators are important tools that help you determine the safety of the atmosphere you are metering.

SAFETY
Because the CGI must combust the sample being drawn through the meter, the oxygen content of the atmosphere must also be between 19.5 percent and 23.5 percent. Oxygen levels above or below that range affect the CGI, and the meter will give an inaccurate reading due to the increased or decreased combustion that will occur.

OXYGEN MONITORING EQUIPMENT

Air normally contains an average of 21 percent oxygen. Lowering or increasing the oxygen content of air can have serious effects on both human life and the fire hazards presented by certain fuels. The initial atmospheric oxygen content within a confined space can be affected by several means and change. If the confined space has been closed for a long time, simple things such as the decomposition of organic materials in the space (for example, leaves, grass, septic waste) through bacterial action could lower the oxygen content of air in the space. The oxygen content in the space might have been lowered intentionally (for example, nitrogen gas for inerting, carbon dioxide for fire suppression) or unintentionally (for example, welding done in the space, painting or coating the interior of the space). Regardless of what caused the oxygen content to be lowered, an oxygen content of 19.5 percent or less by volume in air is to be considered oxygen-deficient, as shown in Figure 3-6. Entry into the space should not be allowed unless the entrant is wearing a positive pressure self-contained breathing apparatus or a positive pressure–supplied air respirator.

Effects of O_2
- Above 23.5%—materials can ignite easily and will burn rapidly
- 23.5%—OSHA definition as oxygen enriched
- 21%—normal
- 19.5%—OSHA definition as oxygen deficient
- 17%—some muscular impairment, increased respiratory rate
- 12%—dizziness, headache, rapid fatigue
- 9%—unconsciousness
- 6%—death within a few minutes

FIGURE 3-6 Oxygen levels that are above or below the normal range of 19.5 percent to 23.5 percent lead to problems that must be addressed.

NOTE
Regardless of how the oxygen content was lowered, any time the oxygen content of a space reaches 19.5 percent or less, the atmosphere is to be considered oxygen-deficient.

Conversely, oxygen-enriched atmospheres—those above 23.5 percent oxygen in air—create their own set of hazards. Though an oxygen level above 23.5 percent will not lead to a person being overcome, it can create a situation where materials normally difficult to ignite or burn can become combustible and burn rapidly. Items such as electrical equipment, high flash point (FP) flammable liquids (FP above 200°F), and clothing are just a few examples of the types of materials that can be affected by an oxygen-enriched atmosphere.

To detect the level of oxygen within an area, use an oxygen meter. Oxygen meters use a sensor that contains an electrolytic material that reacts with oxygen in the ambient air and produces an electric current. Based on the level of oxygen in the air, the sensor will produce more or less electricity, and the change in the electrical current is read by the meter. This reading can be displayed on either a digital readout or an analog gauge and is translated into the percentage of oxygen in air. Unlike the CGI, oxygen meters are fairly straightforward, and the reading is the percentage of oxygen by volume in air. If the meter shows a 20 percent reading, the air within the space being monitored contains 20 percent oxygen. As with the CGI, to correctly monitor the atmosphere within the space, you must monitor at different levels to compensate for the differences in vapor density of different gases.

SPECIFIC GAS MONITORING

The ambient environment in a confined space can include not only combustible gases or abnormal oxygen levels atmospheres, but other gases as well. Some of these gases may be toxic, some may simply displace the oxygen, and some may create both toxic and oxygen-deficient atmospheres. To help you determine the presence of gases and measure how much gas may be present, you need to monitor for specific gases, as shown in Figure 3-7. Gas-specific monitoring instruments are usually designed to detect one or more distinct gases by a chemical reaction between the specific gas and a particular chemical agent in the device. Readings for these devices are given in **parts per million (ppm)**. There are also multi-gas detection devices that simultaneously monitor for combustible gases, oxygen content, and specific gases, as shown in Figure 3-8. These devices contain multiple sensors and provide multiple readouts.

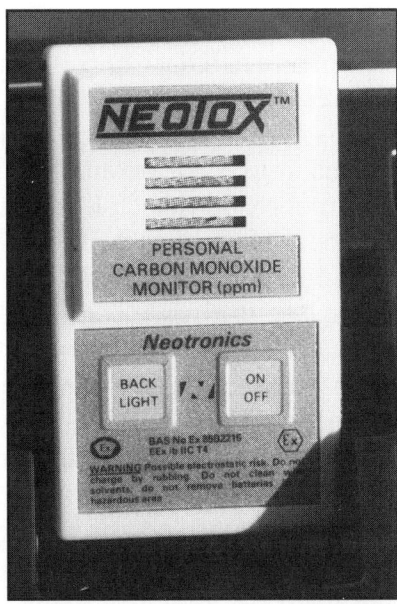

FIGURE 3-7 This is a meter that is designed to monitor carbon monoxide levels. It is not intended to detect or measure any other gas.

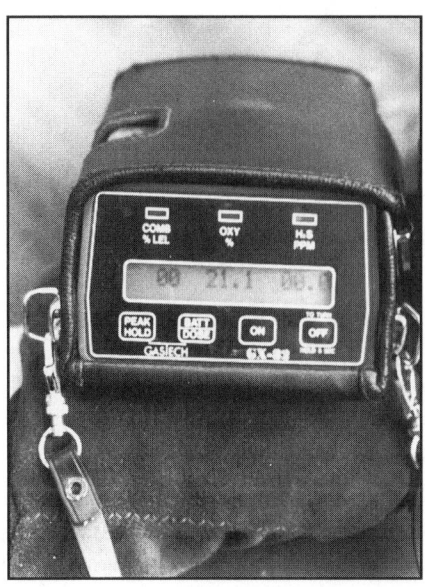

FIGURE 3-8 The meter shown here is designed to measure combustible gases, oxygen, and hydrogen sulfide.

A gas-specific detector such as a carbon dioxide meter is designed to detect and measure carbon dioxide. It does not reliably detect or measure hydrogen sulfide, chlorine, ammonia, or any other gas that may be present in the confined space. This means you must have some idea of which gas you expect to be present or, in the case of a confined space, establish a routine metering protocol that looks for and measures certain gases at all confined space incidents. If you arrive at a confined space incident and are told that the interior of the space was being coated with a paint containing a high level of volatile organic compound (VOC), you will probably not be able to measure the VOC concentration. You may have a meter with you that will measure oxygen levels and combustible gases—and you should use that meter to determine the percentage of LFL and oxygen—but the VOC concentration will remain an unknown and it will become a strategic factor affecting your operations.

As stated previously, gas monitoring equipment provides a reading in ppm. A ppm reading is different than the percentage reading you get from a CGI or oxygen meter. One ppm equals 1/1,000,000 of the whole and is 1/10,000 of 1 percent. This is a significant difference from percentage meter readings. An oxygen reading of 20 percent means that there is 200,000 ppm of oxygen present. A CGI calibrated on methane with its 5 percent LFL would equal 50,000 ppm at a reading of 100 percent LFL.

Why are the gas-specific meters different? Simply because the gases they are designed to detect and measure are toxic at very low concentrations. Carbon monoxide and hydrogen sulfide, among other gases, are considered **immediately dangerous to life and health (IDLH)** at 1,500 ppm and 300 ppm, respectively, well below the threshold that would produce a flammable atmosphere or reduced oxygen content. Neither IDLH limit would be detected by an oxygen meter or CGI. Even though both gases are also flammable—carbon monoxide has an Lower Explosive Limit LEL of 12.5 percent, and hydrogen sulfide has an LEL of 4.3 percent—they do not pose a fire hazard until the CO is at 125,000 ppm (12.5 percent × 10,000) and 43,000 ppm (4.3 percent × 10,000). By the time the combustible gas indicator warned you that there was a fire hazard, it would probably be too late for anyone in the contaminated atmosphere. The victims would have died from the toxic effects of the gas that was dozens of times higher than the IDLH limit. What becomes critical here is to understand that a gas, such as hydrogen sulfide, may have several different properties that make it hazardous in different ways. The gas may be both toxic and flammable, and if you were to use a meter designed for a single purpose, such as a CGI, you would be getting only part of the information you need to develop an action plan for the rescue.

Colorimetric tubes are another type of monitoring device that you may have available to you. The tubes, commonly called Dräger tubes or Sensidyne tubes, are glass tubes filled with a chemical agent that reacts with specific gases or vapors, as shown in Figure 3-9. The gas is drawn into the tube by a pump designed for use with the tubes. When the specified gas is detected, the chemical agent reacts with the gas and turns color. The color change extends up the tube, and by measuring the length of the color change against a scale, usually on the tube, you can determine the concentration of the gas. If your team uses these devices or they are made available for your use at an emergency, you must be aware of the limitations of specific types of tubes. First, these are product-specific tubes, so you must know or suspect the presence of that specific product in order to select the correct

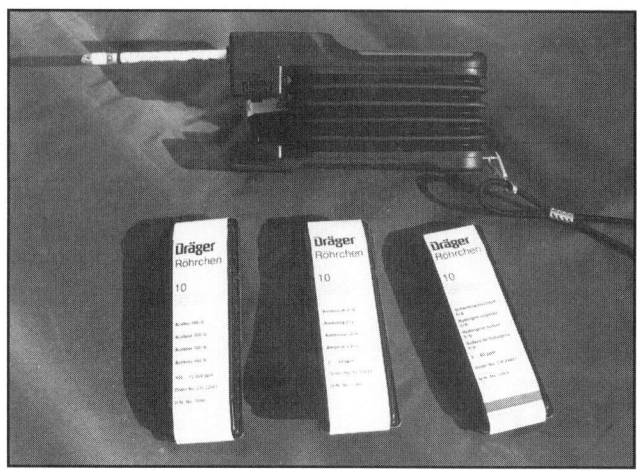

FIGURE 3-9 Colorimetric tubes are designed to be used to detect specific chemicals. These tubes are used by drawing a sample of air through the tube and then noting any color change.

tube. The tube must be used in a pump supplied by the manufacturer, and you must ensure that you draw the proper amount of air through the pump. A manual pump requires you to pump the device using full strokes. In addition, the pump should be in good working condition and not have leaks. It takes time to complete each compression and refill of the pump, and you must accurately count the number strokes. Even with all of that work, the accuracy of your reading is ± 25 percent. This means that a reading of 10 ppm could actually be anywhere between 7.5 ppm and 12.5 ppm. If the gas you are measuring has an IDLH of 11 ppm, is this atmosphere over or below the limit?

Not all gas-specific monitoring devices provide direct readings. For example, a simple home carbon monoxide detector that sounds an alarm when the CO reaches a preset level but does not tell you what the level is. The detector is valuable because it gives a warning to evacuate. Unfortunately, you do not know how much past the alarm point the carbon monoxide level is. This may seem like a simple example and as a trained rescuer you should know the limitations of this type of equipment, but not everyone is properly trained for confined space entry or rescue. There are detection devices on the market that are designed to monitor many different gases and simply sound an alarm when one of those gases is detected. You do not get a direct reading, but only an alarm that the gas is present at or above the preset alarm point. Devices such as these should not be used for confined-space rescue work.

pH DEVICES

In addition to concerns about the flammability or toxicity hazards of materials or the impact the materials may have on the oxygen level, we also need to be concerned about the corrosive character of the material. **Corrosive** materials are classified as either acids or bases (also called alkali or caustic). The pH scale measures the strength of an acid or base. It is a logarithmic scale ranging from 0 to 14. On the pH scale, 7 is neutral. For practical purposes, pH scale is used to measure intensity, not a specific amount of corrosivity. Materials with a pH of less than 7 are acidic, and materials more than 7 are basic, as shown in Figure 3-10. A **logarithmic scale** is based on a multiplier of ten. You need to be aware of this to understand the corrosivity of the material. Each time the pH scale changes by one whole number, the strength of the acid or base increases or decreases by ten times. The strength increases as the pH moves away from 7 (neutral) and it decreases as the pH moves toward 7 from 0. Consider, for example, an acid with a pH of 1.0 compared to an acid with a pH of 2.0. The acid with the pH of 1 would be ten times as strong as the acid with the pH of 2.0.

Materials with a very low or very high pH are strong acids or bases. Contact with these acids or bases can cause damage to human tissue, metals, or other materials because of the intensity of the corrosive properties. To qualify the degree of hazard that you, as an emergency responder, might face from a solid, liquid, or gaseous corrosive, you can determine the pH of the material by using a pH meter, pH pencil, or other pH detection equipment such as **pH paper.** Whatever the method you choose to determine the pH of the material, the detection equipment has to make direct contact with the corrosive material. For that reason, you will need to provide personal protective equipment so that the material does not injure the person using the detection equipment.

pH OF SOME COMMON SUBSTANCES

Substance	pH	
	14	
lye		
	13	
household ammonia	12	
		BASIC
	11	
lime water		
	10	
borax		
	9	
baking soda		
	8	
blood		
milk	7	NEUTRAL
rain	6	
black coffee	5	
tomatoes	4	
soda		
	3	ACIDIC
lemon juice	2	
gastric fluid	1	
	0	

FIGURE 3-10 A pH chart showing the pH range of some common materials. The range is from 0 (very acidic) to 14 (very basic).

> **NOTE**
>
> pH paper is practical to use in the field, but you must know the specific type of pH paper you are using and what a color change indicates.

Meters used to determine pH are simple, direct reading instruments. A **pH pen,** which looks similar to a ballpoint pen, or a more sophisticated pH meter, as shown in Figure 3-11, are two examples of the types of equipment used. The probe is placed directly into the corrosive material, and a reading from 0 to 14 is shown on the instrument readout. Paper strips, like those shown in Figure 3-12, are commonly used. Place the strip in contact with the corrosive

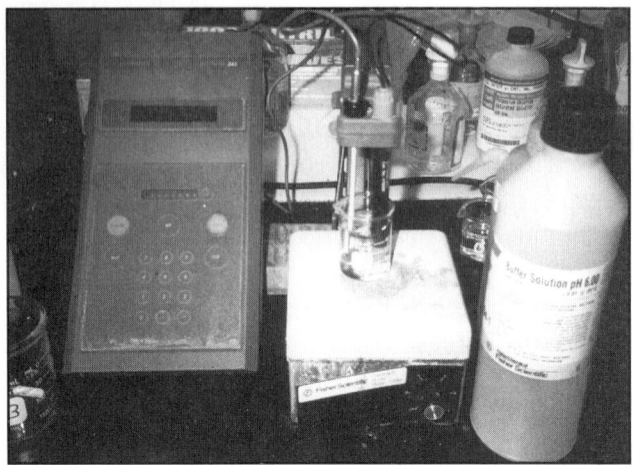

FIGURE 3-11 This device is a pH meter, and it is designed to be used in a laboratory. You could bring a sample of the product to the lab, but in most cases this would not be practical. That is one of the values of a portable meter.

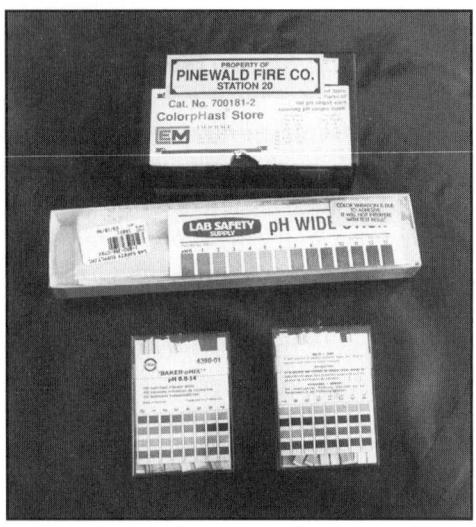

FIGURE 3-12 pH paper is practical to use in the field, but you must know the type of paper you are using and what a color change indicates.

material and then observe the change of color as the material reacts with the agent on the paper. Some of these papers will only turn red or blue to confirm the presence of an acid or base, while others will turn varying shades of colors. Then match the color shade on the strip with a chart that shows the approximate pH. When you are finished using these meters or pH papers, you must be aware that some of the corrosive material can remain on the meter probe or pH paper. Proper handling, decontamination, or disposal will prevent injury to people and the environment. Isolate the equipment until it can be decontaminated or disposed of.

> **NOTE**
>
> After you have used these meters or pH papers, you should be aware that some of the corrosive material can remain on the meter probe or pH paper.

UNDERSTANDING MONITORING EQUIPMENT READINGS

When you use monitoring equipment, there is a distinct order in which you should conduct your testing. To begin, start your monitoring in clean air. This allows you to see how the equipment responds and if it is working. The manufacturer should supply you written instructions with the equipment, and this should be your guideline as to what the normal response of the device is. If you are using separate devices to test the atmosphere, (for example, a CGI, a separate oxygen meter, and a Dräger tube), which should you use first? The proper order for individual testing is to measure the oxygen content and the combustible gases, and then test for toxicity. The oxygen content in the space will affect the performance of the CGI, so measuring oxygen content first will give you an expectation of how your CGI will respond. Of course, if you are using a meter that measures all three items simultaneously, you do not need to follow this testing order. You should still be aware that an oxygen-enriched or oxygen-deficient atmosphere will affect the combustible gas reading, so you should take appropriate steps to monitor, understand, and control potential hazards.

> **NOTE**
>
> If you can not monitor simultaneously, start by measuring the oxygen levels, checking the combustible gas readings, and then checking for and measuring toxicity.

By now you should know that you need to monitor the atmosphere of the confined space, but you also need to know that you should monitor the atmosphere outside of the confined space as you approach. This will serve to protect you from contaminants that may be venting from the confined space and creating an area outside the space with a hazardous atmosphere where rescuers will not expect it. You must always monitor for oxygen content and combustible gases. Depending on what you know about the space, you may also have to monitor for other expected or suspected gases. Preplan with local facilities where you will be the designated rescue team to determine what other gases to expect, what specialized monitoring equipment you will need, and whether this equipment is kept onsite and maintained. Proper evaluation or size-up of the emergency scene will also help you determine what to expect and what monitoring equipment you will need. Your size-up should look for answers to the following questions about the atmospheric hazards of the confined space:

- Is a confined space permit present at the emergency?
- What does it identify as to the type of work being done in the space or what products may have been in the space?
- Are there other workers nearby, and were they briefed about any hazards that might be present?
- Do you see markings or labels on the vessel or tank (that is, the confined space) as to what it contained before entry?
- Did anyone report smelling a distinct odor or taste (you should not confirm this by smelling or tasting) from the product or contaminants in the space?
- What other indications are present to help you identify the atmospheric hazards?

It is normal to expect certain gases in belowground spaces such as sewers and septic tanks or in aboveground spaces such as silos. You combustible gas indicators will tell you how close you are to the LFL of the gas on which the meter is calibrated. Have a plan that identifies what to do if you reach the action limit. Include in your plan a method for controlling the atmospheric hazard. Make sure the instruments have been inspected, tested, and maintained according to the manufacturer's requirements. And do not use a CGI that has not been calibrated, or recalibrated as required.

> **NOTE**
>
> You should monitor the atmosphere outside of the confined space as you approach to be sure that contaminants are not venting from the area and creating a hazardous atmosphere where rescuers will not expect it.

Though oxygen meters will tell you the oxygen content of the atmosphere you are monitoring, make sure you know and understand the impact of oxygen-deficient and oxygen-enriched atmospheres on your operations and equipment. Consider how different types of oxygen atmospheres increase the danger of fire or require the use of respiratory protection.

Specific gas monitoring instruments are designed to detect and read the levels of the gas for which they are intended, but you must also understand that in addition to the specific gases, there are interferants that might affect the meter. Interferants are gases that basically "trick" a meter into believing that the specific gas is present. These interferant gases mimic the reaction the sensors have with the gas the meter was designed for and can give you false or inaccurate readings. A carbon monoxide meter may go into alarm not because it has detected the corresponding level of CO, but because gasoline was present and releasing benzene vapors. The benzene reacts with the CO sensor much the same way that CO does, and it therefore interferes with the operation of the meter. You should know or have access to information that identifies the materials that can cause a false reading on your gas meter before you attempt to use it at an emergency.

Using colorimetric tubes to confirm the presence of a specific material brings real value to your size-up and can be useful in developing your action plan. But beyond that they have some serious limitations, and you must know what those limitations are. The use of colorimetric tubes, like most monitoring instruments used by emergency responders, has been adapted from other fields such as safety and industrial hygiene. Consult with your suppliers and the

SKILLS/PROCEDURES 3-1
Monitoring a Confined Space

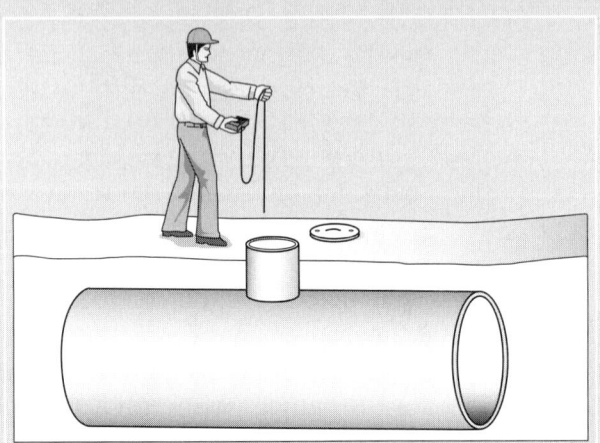

1 Monitor the atmosphere outside of the confined space as you approach the opening to make sure venting contaminants are not creating a hazardous atmosphere that you will walk into. Then begin monitoring the space, beginning at the top of the space, moving to the middle, and finally moving near the bottom of the space.

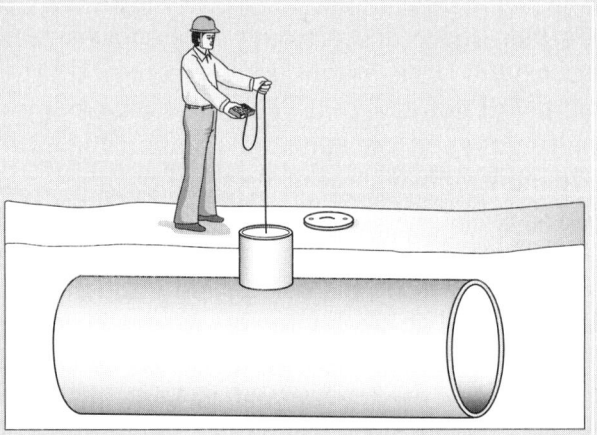

2 Depending on the length of any hose or tubing attached to the monitoring equipment, the response time of the equipment will vary. The longer the hose, the longer it takes for the sample to reach the sensors.

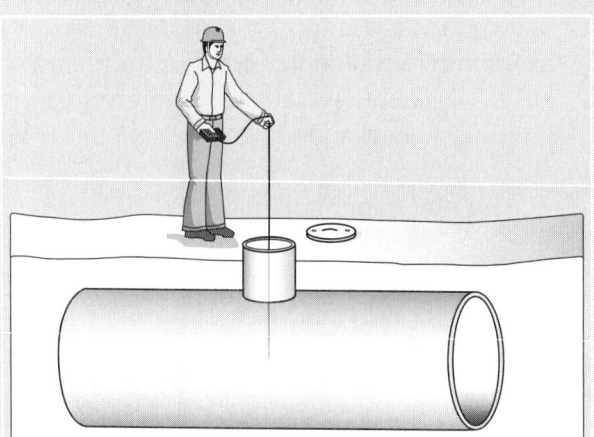

3 If the space is deep, you might want to consider taking readings at intermediate levels between the top, middle, and bottom levels.

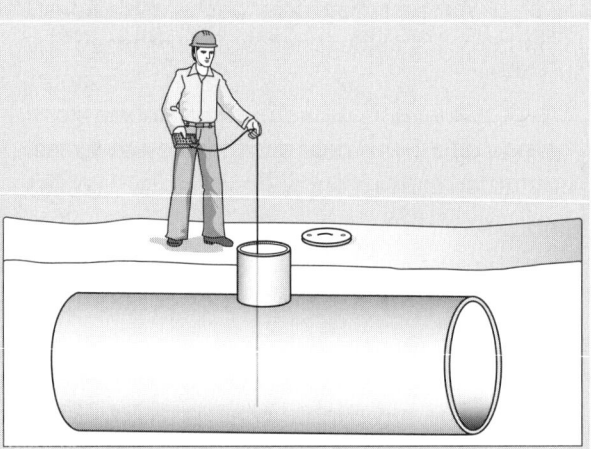

4 Do not allow the sampling tube to touch or rest on the bottom of the confined space. Liquids, dirt, and other debris can enter the sampling hose and plug or damage the meter.

manufacturer of the colorimetric tubes so that you understand the limitations and capabilities of the specific equipment you have.

You should also be aware of other items that can affect the operation of monitoring equipment. These items that need to be addressed include **radio frequency (RF) interference,** proper setup of the instruments, manufacturers' instructions and limitations, actively monitoring the atmosphere at all portions (elevation and enclosures) of the space, calibration, and training. Using portable radios in close proximity to some electronic monitoring devices can cause a false alarm. You should not transmit on a radio that is close to a monitoring instrument. Keep the radio and the monitoring device separate to avoid RF interference. Check with the manufacturer to see if this is a known problem. When you are going to use a monitoring instrument, it should be set up (turned on, warmed up, and zeroed, if needed) in fresh air. This fresh air setup should be regarded as the acceptable starting point and helps avoid using a contaminated background reading for your starting point. Read and follow all manufacturers' instructions. The manufacturers know the ability and limits of their equipment better than you; they are liable for the equipment's performance and want to protect you from doing something foolish with it. To compensate for not all gases having the same vapor density, you need to monitor at all levels of the confined space and, if there are enclosures within the space, those areas as well. Checking only the top or bottom of an area measures only that portion of the space and can provide you with a false reading. Remember that materials that are lighter than air rise and accumulate at the top of a confined space, and materials that are heavier than air sink and accumulate at the bottom. Monitoring at only one level can cause you to miss a high concentration of an atmospheric hazard. Monitor at all levels—top, bottom, and intermediate—as shown in Skills/Procedures Box 3-1. Divide the vertical levels of the space so that you break it up into multiple levels about every 5 feet. Allow enough time between each sample for the atmosphere to enter the tube and be drawn into the meter, and for the meter to provide a reading. Skipping levels or moving the sampling tube too quickly will not give you accurate readings.

Contact the manufacturer or supplier of your monitoring equipment and see what training classes or information they can provide. Regular calibration of the monitoring equipment is critical to proper performance. An uncalibrated monitoring instrument might give you false high readings and delay your entry and rescue of the victim. Worse than a high reading is a false low readings or no reading at all. Do you then assume that all persons on the rescue team are safe when in fact they are in serious danger? Not training with the monitoring instruments before you need to use them in the field is almost as bad as not calibrating the equipment. Proper training is just as important as calibration. Identify and understand the limitations of your equipment. Know what the meter reading means and how you can use that information. That knowledge lets you protect both the rescuers and victims.

LESSONS LEARNED REVISITED

Rescue crews found no oxygen in the space, and the autopsy report identified the cause of death for both victims as "asphyxia due to air displacement/hypoxia due to nitrogen gas."[1]

Investigation by both the local sheriff's office and the meter manufacturer showed that the multi-gas meter was functioning properly.

Discussion Questions

1. If you were the confined space entry supervisor, would you have required the use of a harness and retrieval equipment for attendant-based rescue?
2. The foreman reported the atmospheric reading as "fine." Should you accept that definition, or should you ask for the specific readings?
3. The car originally contained soybean oil and only the headspace was filled with nitrogen. The rail car had been off-loaded prior to being shipped to the yard, and air replaced the soybean oil. How would you monitor this car for oxygen content if the tank was 10 feet high and the hose for your O_2 meter was 8 feet long?

---

[1] "Two Railroad Repair Workers Asphyxiated in Damaged Tank Car" Iowa Face Report Case No.: 01IA021 Report Date: July 24, 2003.

SUMMARY

- Monitoring instruments enable you to characterize the atmospheric conditions at a confined space incident.
- You can obtain accurate information and make the greatest use of that information if you know:
  - The detection capabilities and limitations of the instruments
  - How to translate the data produced by the instrument into usable information
  - What that information will mean to your rescue operation
- Equipment limitations include the accuracy of the readings and the quality of the routine maintenance program (for example, calibration).
- Understand and follow the manufacturer's instructions.
- Properly maintaining and operating your monitoring equipment will allow you to begin to trust the readings you get from the equipment.
- Failing to maintain the monitoring equipment can produce readings that are wrong—dead wrong.
- Set up your equipment in a clear area to avoid getting a false reading from contaminated air.
- Monitor at all levels of the confined space and continuously monitor as long as an atmospheric hazard has the potential to affect your operations.
- Develop procedures to guide you through the proper steps required to use monitoring equipment correctly.
- Practice with your monitoring equipment and learn how to use it long before you need to use it at a confined space incident.
- Training and practice increase your proficiency with the monitoring equipment.

REVIEW QUESTIONS

1. You are using a combustible gas indicator and have a meter reading of 8 percent. What does this reading mean? Is there an 8 percent concentration of combustible gas in air, or is the atmosphere at 8 percent of the lower flammable limit?
2. While using an oxygen meter, you get a reading of 28 percent. Does this mean you have an oxygen-deficient, oxygen-enriched, or normal oxygen atmosphere? What is the most critical hazard you face from this atmosphere?
3. A multiple gas detector (combustible gas, carbon monoxide, hydrogen sulfide, and oxygen) is producing a reading of 15 for carbon monoxide. What unit of measure goes with this reading: percentage of gas in air, percentage of the LFL, or parts per million?
4. A pH meter reads 1. Would this reading indicate an acid or a base? Would the reading indicate a stronger or weaker intensity than a pH of 2?
5. Combustible gas indicators tell you the percentage of combustible gas in air. True or false?
6. The CGI is meant to detect the ____ of the gas with which it is calibrated.
7. What levels of a confined space should you monitor before entry?
   a. top
   b. bottom
   c. middle
   d. all of the above
8. What does IDLH mean?

KEY TERMS

direct reading instruments 32
Combustible Gas Indicator (CGI) 32
Action limit 34
parts per million (ppm) 37
Immediately Dangerous to Life and Health (IDLH) 38
Corrosive 39
Logarithmic scale 39
pH paper 39
pH pen 39
Radio Frequency (RF) interference 43

ACTIVITIES

1. Review the manufacturer's instructions and practice using your monitoring instruments. Depending on the particular device, how it is meant to be used, and what it is designed to detect, you may be able to use it with both known and unknown products. Examples of this include:

   - Using a combustible gas indicator with calibration gas samples obtained from the manufacturer. The quantity of gas in air is known and you may be able to obtain several different gases in order to show how the different LFL of each gas produces different readings on the same instrument.
   - Using a CGI with a sample of a flammable or combustible liquid. Use caution if you do this. Wear proper protective equipment, limit the quantities that are present, make sure the area is ventilated, make sure it is safe to handle the materials in the space, and make sure that the product will not damage your CGI.
   - Using pH equipment with simple chemicals such as coffee or vinegar. Ensure you wear the proper PPE, handle and dispose of the material properly, and decontaminate or dispose of the pH equipment properly.
   - Setting up and operating your monitoring equipment according to your procedures.

2. Ask a manufacturer's representative to provide a demonstration and training on the proper use of the equipment.

ADDITIONAL RESOURCES

For additional information on the topics discussed in this chapter, check out the following:

American Society of Safety Engineers, http://www.asse.org/

American Institute of Industrial Hygienists, http://www.aiha.org/Content

Don't forget about manufacturers' and vendors' websites. They can also have valuable information.

Local industrial safety and health departments that use monitoring instruments are also great resources.

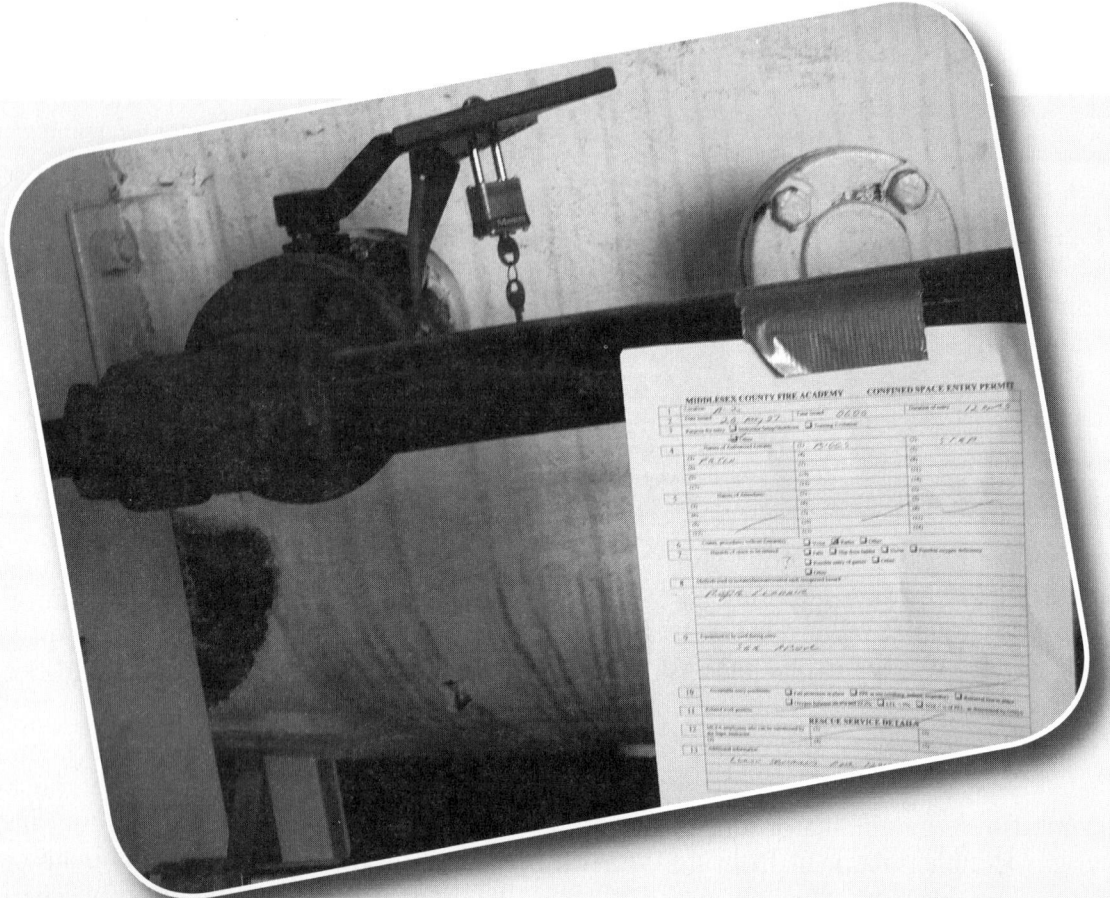

4 Lockout/Tagout

LESSONS LEARNED: TWO WORKERS CAUGHT IN RUPTURE OF PRESSURE VESSEL

Two workers entered the lower stage of a two-stage pressure vessel to clean the interior and perform maintenance on the vessel. Both stages of the pressure vessel had design pressures of 3 psi and normally ran at 6–8 psi. The entire unit had been purged with nitrogen earlier in the day to remove toxic gases from the vessel. During this purging, the pressure spiked to over 20 psi in the upper stage. Several hours later, the lower stage was deemed safe to enter and workers entered to begin their work. Shortly after entering, the upper stage ruptured.

Critical Thinking Questions

1. What forms of energy can you identify as being involved in this emergency?
2. If you were called to a similar emergency, what could you use to identify the forms of energy involved in the accident?
3. If you identified over pressurization of the upper stage as the direct cause of the failure, what should you consider doing to secure the pressure source?

LEARNING OBJECTIVES

NFPA 1006, *2008 EDITION*; OSHA 1910.147
By the end of this chapter, you should be able to:

- Control hazards (NFPA 1006 7.2.3):
  - Establish a rescue area.
  - Protect rescuers and victims from the hazards of the confined space, including:
    - Exposure to hazardous materials
    - Hazardous atmospheres
    - Stored energy releases
    - Additional harm to victims
- Manage potential energy sources, by evaluating potential energy sources to identify (NFPA 1006 10.1.5):
  - Potential energy hazards to rescuers and victims
  - Unsafe release of potential energy
  - Means of control for all energy systems
  - Beneficial use during the incident

Identify confined space operations where *unexpected energization, up of the machines or equipment, or release of stored energy could cause injury* to rescuers (OSHA 1910.147).

- Identify the following energy sources:
  - Electrical
  - Hydraulic
  - Pneumatic
  - Chemical
  - Thermal
  - Gravity
- Define how preplanning can assist in identifying the need for lockout/tagout.
- Define how a hazard and risk assessment should include a review of lockout/tagout requirements.
- Identify examples of equipment and methods that can be used for lockout/tagout.

INTRODUCTION

As you are looking for something in your kitchen cabinets, you see evidence that a mouse has been entering your home and eating food stored in the cabinet. You carefully bait a mouse trap and leave it in the cabinet overnight. In the morning, you check the trap and find that the mouse has eaten all of the bait, but the trap did not release and catch the mouse. As you reach down to pick up the trap and re-bait it, the spring releases and the bar narrowly misses catching your finger. What happened, and how does this relate to a confined space emergency?

When you baited the trap the night before, you pulled the bar back against a spring and locked the bar in place with the releasing mechanism. The spring wanted to return to its original position but could not. You stored all of the energy in the spring, which created potential energy. When the releasing device activated, the stored energy became kinetic energy and brought the bar (powered by the spring) back to its original position. Had your finger been in the way of the bar, it would have absorbed the energy created by the spring. At the very least, you would have had a sore finger from getting struck by the bar.

Let's take that situation to a confined space incident where several hundred pounds of grain are stored in a cone-shaped hopper. The victim is in the hopper, surrounded by the grain. This person is trapped at the bottom of the hopper with their head and arms visible inside the hopper, and their legs can be seen hanging out of an opening at the bottom of the hopper. The potential energy source here is gravity, and it is stored in the grain. The grain wants to flow down and out of the hopper. It is already pressing on the victim who is, in essence, a stopper in the opening. If you can remove the grain or somehow keep it from flowing through the opening, you can control the energy source. But what would happen if you put a rescuer on a ladder under the victim and had them push up on the victim? As the victim came out of the opening, the grain would be free to flow. Would it run down on the rescuer and bury her, or would it knock her off the ladder and injure her? The key here is to identify the potential energy source and control it so that it does not endanger either the rescuer or the victim.

FIGURE 4-1 A lockout control center showing some of the equipment required for controlling energy sources.

> **NOTE**
> The unexpected energization or startup of machines or equipment, or the release of stored energy, can cause injury to people.

To understand potential energy sources, you need to begin by understanding the hazards that those sources present. OSHA Standard 1910.147 (Lockout/Tagout) defines an energy source as "[a]ny source of electrical, mechanical, hydraulic, pneumatic, chemical, thermal, or other energy." This is a broad definition, but it is where you should begin, using some basic examples of what is an energy source. If you entered a room marked "Danger High Voltage," you would expect to find electrical equipment and would at least be aware that an electrocution hazard existed. But do you recognize the potential contained in a reactor vessel with an agitator paddle or a large storage tank with pipe lines running into it? Would you anticipate the potential start-up of the paddle while you were in the reactor? Or that a valve could be opened (sometimes remotely) and product would flow into the storage tank?

Imagine that your rescue team has entered a 250,000-gallon fire protection water tank to rescue an unconscious victim. At the base of the water tank is a manually operated valve on a 12-inch water feed line. The rescue entry was made from an opening on the top of the tank, and as you prepare to remove the victim, a remote valve on the 12-inch feed line is opened by a plant operator located in a control room hundreds of feet away from the incident. Depending on whether the manually operated valve is open or closed, what is going to happen? Now visualize that, instead of water, the product is a flammable, toxic, or corrosive liquid. Does everyone inside the tank drown? Do they suffer burns from the corrosive liquid? Or are they overcome by the fumes of the toxic liquid? In this case, the energy in the liquid presents a physical hazard accompanied by the chemical properties of the product. The key to controlling potentially hazardous energy sources is to first identify the problem and then take control of the energy sources, as shown in Figure 4-1. Identifying the intended use of the space, the contents of the space (for example, energy, product), and how the energy is delivered to the space are indicators of both the potential hazard and the methods of controlling that hazard.

If you were an incident commander, would you blindly commit rescuers to a confined space with live electrical equipment, a hydraulic cylinder, a pressurized air line, or an unsecured heavy metal lid? How you handle the incident will depend on your ability to either control or eliminate potential hazards. At times you will be able to remove the hazard, but at other times your only option may be to get control of the hazard. So where do you begin?

> **NOTE**
> You will need to seek out the confined space entry permit, the entry supervisor, the attendant, and any other person who witnessed or has knowledge of the incident.

LOCKOUT/TAGOUT REQUIREMENTS

Ideally, your hazard recognition process should begin with preplanning, onsite visits, and onsite training. If you work for a private company and are part of their onsite confined space rescue team, it is much easier to survey confined spaces at your facility. In practice, identifying the lockout/tagout requirements should be part of the confined space entry permit issuing process. You may even want to take the confined space entry process a step further by identifying the hazards and the methods of control in the event of an emergency. This could be in the form of an onsite preplan update for a specific confined space entry. If you are part of a public sector response team, instead of a private industry team, you must still consider some level of incident preplanning, especially if your team has agreed to be the designated rescue team for a particular location or facility. You should also look to either use similar confined spaces for training at the facility or train by conducting simulated rescues based on the features of the confined space.

Preplanning

Preplanning provides you with an assessment of the situation before an emergency. It should be an accurate and in-depth size-up of potential emergencies and should identify the basic strategic elements that you will be faced with during an actual emergency. Before the incident, you can focus on key elements without the sense of urgency you would feel during an emergency. You can identify the hazards of the space, the means of controlling those hazards, and the resources you will need to have available during an emergency. Among those hazards and methods of control should be potential energy sources, the methods used to control them, and how to confirm that the energy has been isolated. You should be able to identify the action options needed for the anticipated emergencies and craft a basic action plan to initiate your rescue operations.

A word of caution is appropriate here. Even with a comprehensive preplan, you cannot take for granted that conditions in the space have not changed since your initial size-up. Even with your last preplan inspection only hours old, conditions within or near the space may have changed. Your preplan should also be considered a checklist to compare what you know about the facility with the conditions that exist at the time of the emergency.

So far, we have been talking about preplanning, but what if there is no preplan for the confined space accident you have been called to? Whether or not there has been any preplanning, you must conduct a size-up of the incident. The size-up is intended to identify the limiting or strategic factors related to the incident conditions. Obviously a preplan should have already identified those factors, but with or without a preplan, you must be prepared to conduct a **hazard and risk assessment.** Where do you start?

- Develop a method to determine the accuracy of the information you are receiving.
- Identify potential energy sources.
- Determine which energy sources require lockout, as shown in Figure 4-2.
- Identify and control sources of residual or stored energy such as a compressed spring or a charged electrical capacitor.

FIGURE 4-2 Product stored in this hopper is intended to be released through the chute at the bottom. The source of the energy moving this product is gravity, and it must be thought of as stored energy.

> **NOTE**
> As part of your confined space response equipment you should carry basic lockout/tagout equipment.

> **NOTE**
> Identify all potential energy sources, determine whether they are hazardous energy sources or **beneficial energy sources**, and take control of them.

> **NOTE**
> Beneficial energy sources are those forms of energy that you will use for assistance during an emergency (for example, electricity for lights and ventilation equipment, natural gas for emergency generators). Beneficial energy sources should not be allowed to pose a hazard during the incident. It may be necessary to take steps to control the hazards so that the energy can be safely used.

Hazard and Risk Assessment

This chapter is devoted to lockout/tagout. We will limit any discussion of hazard and risk assessment to hazardous energy sources and determining how they will affect your operation and the safety of personnel. You should start your hazard and risk assessment with three simple questions: What has happened? What is happening now? The third question is based on the answers to the first two questions: what can you predict will happen as this incident plays out?

When you arrive at any emergency, you need to determine what has happened. EMS personnel look for the mechanism of injury at an accident scene. Firefighters need to locate the seat of a fire. As a member of the confined space rescue team, you will do the same thing, and you need to see the whole picture. That includes the victim, the confined space, and the area surrounding it. Seek out the confined space entry permit and then find the entry supervisor, the attendant, and any other person who witnessed or has knowledge of the incident. These are all pieces of the puzzle and you must figure out how they fit together. To determine what has happened, you should answer the following questions:

- Are energy sources or mechanical equipment involved or potentially involved in the incident?
- What effect will the energy source have on your rescue operation?
- Can you control the energy source, or is it adequately controlled already?
- If the victim is trapped in the equipment or energy source, do you have the necessary skills, training, and equipment to rescue him?
- Will you need to call for additional help from a team that is specifically trained in machinery rescue?

The second question—What is happening now?—is critical to your size-up. If an energy source was involved in the incident:

- Has the equipment been de-energized, does it still contain stored energy, or is the answer that no one knows for sure?
- Is this a beneficial energy source that must be kept in service to support the rescue?
- Can the energy sources be controlled from outside the space, or is entry required?
- How is the energy source affecting the victim or the potential for rescue?
- Can you verify that the energy source has been controlled or de-energized, and how the control is to be implemented?
- Is this a rescue or a victim recovery?

If it obvious that your victim is dead from contact with an energy source (for example, crushing, electrocution, massive thermal burns from steam), your entire operation should change from a rescue to a recovery. Protecting the emergency responders must become paramount, and you may need to take additional steps to protect them. You must compare what is being risked to what you expect to accomplish. If little can be accomplished (recovery), little should be risked. If much is to be accomplished (rescue), how much are you willing to risk?

The third question—What can you predict will happen as this incident plays out?—may seem a little like fortune-telling, but it is not. Your experience and the depth of your knowledge will help you make fairly sound predictions.

If there are open electric circuits in the confined space, the space is very narrow, and your rescuers have to enter wearing Self Contained Breathing Apparatus (SCBA), you can predict, with some certainty, that there is a good chance someone will make contact with the electrical circuit and may be electrocuted. Similarly, you would expect to find crushing injuries during your assessment of a victim trapped beneath a heavy object. These are simple predictions based on knowledge and experience, but what happens when you have little or no information? There will be times when your decisions and course of action may be based on a best guess as to the outcome: You might find that the information you have regarding the emergency leads you to anticipate one thing will occur only to have something completely different happen.

So far this chapter has discussed the need for lockout/tagout. Now we need to turn the discussion to some of the methods used to of control energy sources. With lockout/tagout, you physically control the energy source by stopping the flow of energy. You can lock a valve in the proper position (yes, some might have to be locked open to redirect the energy or relive pressure), open and lock an electrical switch, or provide some other energy-controlling device. Once you have locked the device, you tag it to identify the device and to advise others that the device has been isolated as a safety measure. The goal is not just to isolate the energy source but to prevent others from removing the isolating devices and operating the energy source. Your program must stipulate that no one attempts to remove the lock and tag other than the person who placed the lock or device.

Lockout/Tagout Devices

The types of devices that can be locked include **manually operated electrical circuit breakers, disconnect switches,** and **line valves**. These devices can be locked out using specially designed lockout/tagout equipment, as shown in Figure 4-3. Other devices for lockout/tagout include **latches, chains**, and **chocks** that are used to secure the energy sources. It is also possible to block the flow of product by inserting **blank flanges, blocks,** and **bolted slip blinds** in a pipe, as shown in Figure 4-4.

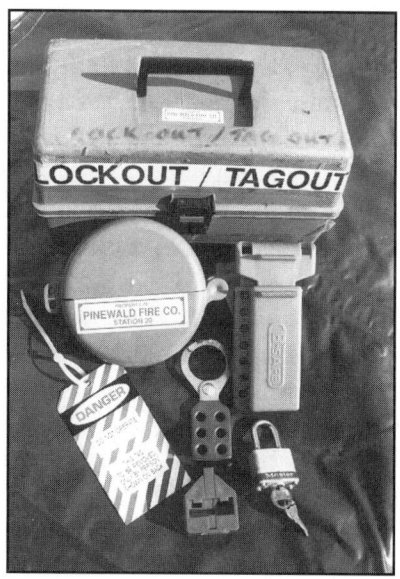

FIGURE 4-3 Lockout/tagout equipment is intended to control various types of equipment. Shown here are some locking devices for valves, switches, and electrical equipment. The confined space rescue team should have basic lockout/tagout equipment as part of their equipment.

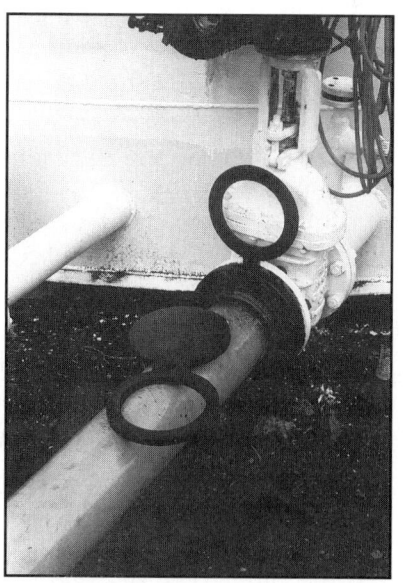

FIGURE 4-4 The device shown in this photo is called a spectacle blind because it resembles a pair of spectacles. One side of the device is solid (blind) and is intended to stop the flow of product through the pipe while the other side is open.

Circuit breakers and disconnect switches are used to isolate equipment from electrical sources. Most of us are familiar with circuit breakers in our homes or where we work. The circuit breaker controls a single circuit that may have multiple electrical equipment attached (think of your home and how many rooms were affected the last time a breaker "blew"). Disconnect switches are often attached to a single piece of equipment and are intended to isolate that piece of equipment from the electrical sources. However you isolate electrical equipment from its source, you must test the circuit or equipment to ensure it is in fact de-energized. Certain types of electrical equipment, especially critical equipment, may have more than one power source or a backup power source.

Line valves, blank flanges, blocks, and bolted slip blinds are all devices that are intentionally inserted in a piping system to control the flow of gases, fluids, or other flowable products. The line valves are usually normal control valves (think of the valves on your sink's faucet) that are used to control the flow. When line valves are used for lockout/tagout, they can be re-opened if they are not physically secured to prevent operation. Blank flanges, blocks, and bolted slip blinds usually require that the piping be disassembled (also called broken) at a flange or connection and the lockout tagout device be inserted into the line. These flanges, blocks, and blinds are more definitive lockout/tagout devices because they must be removed before product can flow through the line, but using them also requires more work, time, and expense.

Latches, chains, and chocks are used to secure equipment or devices that contain energy sources (such as chaining a line valve in the closed position). You can latch a piece of equipment in place to prevent movement if the latch is designed for that purpose and is working appropriately. You could also place chocks (chocking a vehicle's tires is a good example) to stop movement. Whether you use latches, chains, chocks, or some other type of equipment, make sure it is appropriate for the energy source, can withstand the load that will be placed on it, and cannot be tampered with.

Whatever type of device you choose for lockout/tagout, keep in mind what it is that you are trying to achieve. Your goal is to control or eliminate a potentially hazardous energy source, you want to protect the rescuers and victims, and the device must be effective and not easily bypassed, as shown in Figure 4-5. From time to time, you may be tempted to station a person at the switch, valve, or other control as your lockout/tagout method. Using a person instead of a mechanical device is not an acceptable lockout/tagout method. Mechanical devices don't get called away from the location, they don't open because of miscommunication, and they don't have to think about what they are doing. Mechanical devices either work or they don't work, and you can check the performance of the device to ensure it is working properly. If there is no other means available for lockout/tagout and feel you must use a person as your lockout/tagout device, it must be considered a method of last resort, done under only the most extreme conditions and strictly managed.

NOTE

As part of your confined space response equipment, you should carry some basic lockout/tagout equipment.

FIGURE 4-5 This pipeline is locked out by breaking the line and placing a blind flange on the open end of the pipeline.

Lockout/Tagout Equipment

To be properly prepared for confined space rescue, you should include some basic lockout/tagout equipment as part of the inventory of equipment you carry. You can identify the need for specialized or more sophisticated equipment as you preplan sites. During preplanning, you should also determine what specialized equipment may be on-site. With or without specialized equipment, if your basic equipment cannot do the job of lockout/tagout, look at improvised methods such as disconnecting and misaligning pipelines and removing drive belts, chain drives, and mechanical linkages. These improvised methods may be just as effective as specialized equipment, and it may be possible to accomplish lockout/tagout in a very short time. However, using improvised methods requires detailed knowledge of the facility, equipment, and processes.

> **NOTE**
> Disconnecting and misaligning pipelines, belts, chain drives, and mechanical linkages are acceptable forms of lockout/tagout.

Lockout/Tagout Strategic Factors

Earlier in this chapter, we discussed strategic factors. Strategic factors are limiting factors that affect how you handle an emergency. Examples of these factors include the number of victims, the location of the space's access point, the location and condition of the victim, and many other conditions that influence the rescue options available to you. These factors require that that you either change the factors (for example, venting the space to provide air to the victims) or work within the limits set by these factors (for example, the number of victims). Even as you take the limiting factors into account during your size-up, you still must answer those three same questions: What has happened? What is happening now? What is expected to happen? If you find an energy source is present, you must determine how it will affect your operation:

- Is the energy source part of the cause of the accident?
- If it has been locked out, was it locked out adequately?

FIGURE 4-6 Warning sign indicating the presence of automatic starting equipment. Automatic starting equipment can start without warning.

- Is that lock still in place and functioning?
- Does your preplan for this location identify the lockout/tagout requirements and methods?
- Does your preplan anticipate the energy sources and provides a basic action plan?
- Is there equipment that automatically starts, such as pumps and boilers, as shown in Figure 4-6?
- Is your plan up to date, or have conditions changed?
- If there is no preplan for the facility, where do you start?

With or without a preplan, start by looking at the situation you are facing. The preplan will speed up the size-up process because you should already have some verifiable information. But you still need to confirm that information. When potential energy sources are found, look at the type of energy, the source, knowledgeable on-site personnel, the available controls, your own equipment and its limitations, and your standard operating procedures (SOPs). You should then determine whether you have the capability to deal with the conditions with which you are faced. If you can control the energy (or you find it is not a problem), you can develop your action plan to rescue the victim. If you determine that you come up short and cannot control the energy or its source, you may need to reach out for help from others who can control the energy source. That call for assistance should then become part of your plan, and you should consider what actions you can take to either protect lives or stabilize the incident until additional help arrives. Very few agencies have the ability to handle every emergency on their own. Something as simple as live electrical power lines often requires specialized assistance from people outside of the emergency services. That is one more value of preplanning. You recognize the need for specialized help and know how to secure those additional resources.

> **NOTE**
> Look at the whole picture. Do not allow tunnel vision to occur.

As emergency responders, we may not initially recognize the need for lockout/tagout. We may not recognize the presence of potentially dangerous energy sources, or we may be used to addressing items such as utilities control at a later stage during an emergency. This initial emphasis on lockout/tagout can seem confusing, but it demands our attention. A confined space may hold the potential for hazardous atmospheres, engulfment, or energy hazards such as mechanical or electrical equipment, or the space may be intended to store gases, solids, or liquids. Confined space rescue requires you to look at the whole picture. Do not get tunnel vision and focus only on the obvious (the victim, the opening, the way in and out). Look around the confined space and the area it is in, and see if pipes, electrical lines, hoses, or equipment enters the space. Look to see what is attached to the space, including temporary devices and equipment. Determine if energy sources are present and if they are involved or can become involved (especially automatic start up equipment), and then decide if there is a need for lockout/tagout. Under ideal conditions, you can quickly develop all the information you will need to identify and control hazardous energy sources, but emergencies do not occur under ideal conditions. Develop the needed information as quickly and accurately as possible. Verify the accuracy of your information. Question the people who witnessed the accident as well as those working at the site. Failure to control potential energy sources can kill or permanently disable people, including members of your team. Identify the problem and manage the risk with appropriate lockout/tagout controls. You may have to release the stored energy or otherwise control it before attempting a rescue, but you will not know that until you actively look for the hazard and evaluate it.

> **NOTE**
> When you are faced with little information, you must begin developing that information as quickly and accurately as possible, which means that you will have to question anyone who witnessed the accident and those working at the site. Failing to control these energy sources can kill or permanently disable your personnel.

LESSONS LEARNED REVISITED

One worker inside the lower stage was crushed by an internal dome. The second worker was able to escape and survived. A third worker standing on the platform outside the entry point was blown off the platform, through a metal guardrail, and fell to his death.

Discussion Questions

1. If you identified other sources of energy as being involved in this emergency, how would you control them during your emergency operations?
2. Because the upper stage was pressurized with nitrogen, would you anticipate that a low oxygen atmosphere is present?
3. Is you confined space emergency a rescue or recovery at this point?

SUMMARY

- Confined spaces can contain physical hazards.
- You must be aware of all the hazards within the space.
- Identify existing and potential energy sources.
- Eliminate or control potentially hazardous energy sources.
- Physically lock out the energy source.
- Tag the lockout devices to keep track of them.
- Preplanning can identify potential hazards as well as beneficial energy sources.

REVIEW QUESTIONS

1. List at least four potential energy sources that you might come across in a confined space. For each source you identify, explain why you would expect to find it in that space and identify if it can endanger rescuers.
2. Given a facility to preplan for confined space operations, categorize what you would want to identify for lockout/tagout purposes. Who might you consider as a qualified person to answer questions about the facility's lockout/tagout program and equipment?
3. Hazard and risk assessment requires you to identify what has happened and what is happening now, and then predict what you expect to occur as the incident plays out. If you can make an initial determination of the cause of the accident and you can see what is taking place as you size up the incident, how will this allow you to create an energy control action plan?
4. Identify four basic means to accomplish lockout/tagout. Explain where they might be used and why one is better than the other.

KEY TERMS

hazard and risk assessment 49
beneficial energy sources 50
manually operated electrical
  circuit breakers 51
disconnect switches 51
line valves 51

latches 51
chains 51
chocks 51
blank flanges 51
blocks 51
bolted slip blinds 51

ACTIVITIES

As part of preplanning confined spaces within your response area, identify lockout/tagout requirements for the identified spaces and identify the lockout tagout program designed to address these requirements. Speak with facility management to find out if they can show you how the equipment is used and whether they can assist you with training for the equipment.

ADDITIONAL RESOURCES

For more information on the topics discussed with this chapter, check out the following:

OSHA representatives, http://www.osha.gov/

NIOSH, http://www.cdc.gov/niosh/

The State Department of Labor

National and local chapters of the following organizations:

- American Society of Safety Engineers, http://www.asse.org/
- American Institute of Industrial Hygienists, http://www.aiha.org/Content

Local facilities, including:

- Water treatment plants
- Facilities where you are part of the designated rescue team
- Colleges and universities with health and safety programs

5 Using the Incident Command System

LESSONS LEARNED: TWO FIREFIGHTERS TRAPPED IN THE COLLAPSE OF A BURNING BUILDING

Firefighters responded to a working fire in a large commercial building containing five stores early on a Sunday morning in May. Approximately 25 minutes into this multiple alarm fire, the roof collapsed, trapping two firefighters. One of the trapped firefighters was operating a hose line inside the burning store while the other firefighter was ventilating the roof when the collapse occurred. Rescue efforts were started immediately, but the initial headcount showed only one firefighter to be missing.

Critical Thinking Questions

1. How does an incident management system provide structure and coordination to emergency operations?
2. What is meant by the following statement? The incident management system should be flexible and based on the needs of the incident.
3. How does resource accountability support emergency operations?

LEARNING OBJECTIVES

NFPA 1561 *Standard on Emergency Services Incident Management System, 2008 Edition*
By the end of this chapter, you should be able to identify:

- The value of adopting and using an incident management system for all emergency incidents
- The value of a written incident management system
- The application of the incident management system to match the requirements of different types of incidents
- The expansion of the incident management systems to meet operational needs based on the size and complexity of the incident

INTRODUCTION

You are having a house built and are acting as the general contractor. There are masons, carpenters, plumbers, electricians, and other building trades that you have hired. The masons want to begin by building the foundation, but the carpenters insist that they have to leave the job site in a couple of days and would prefer to build the frame first and then lift it on the foundation. The plumbers can't be back to the job site before the insulation and sheet rock are to be installed, and the electricians don't have the supplies they need to wire the house. What type of house do you think will be built under these conditions? How do you manage the different tasks that need to be performed, and how do you place them in the proper order so that, at the end of the project, you have an acceptable house to live in?

Now imagine that you are at a confined space incident. The production manager wants to start production again as soon as possible, EMS would like to get to the victim as soon as they can to treat the victim, the rescue company is responding from another incident 3 miles away, and all of your confined space rescue equipment is with the rescue company. On top of all that, a facility worker is trying to enter the space to assist his fellow worker in the space, and you have asked the police to restrain him from entering the space. How many different groups and how many different agendas are in place under these conditions? You need to put all these different interests in perspective and either convince them to work together or move them away from the incident scene. There are at least three **emergency services organizations (ESOs)** at this incident (fire, EMS, and police) that must be part of the operation, and there is also facility management and employees. You want the support of the facility manager, you may need to call on plant employees to provide technical support, and there may be an onsite emergency response team.

The key to managing any emergency is to have a system in place designed to provide both a management structure and a process for coordination of all the incident resources. The structure should be recognizable so that functional responsibilities are clearly identified and the lines of communication and authority are unimpeded. The structure must be flexible so that it can grow or shrink with the needs of the incident. Implementing the management system at all emergencies makes the use of the system a standard practice for your emergency operations. This is the basis for using an **incident management system (IMS)**.

Beyond providing a management structure for your emergency operations, there are other reasons for using an **incident command system (ICS)** or the IMS. You will need to create an incident action plan that includes addressing risk (people and property), accounting for your resources (people and property again), providing unity of command, and creating an effective span of control. The goal of the action plan is not to just get the job done, but to know who is where, doing what, and when. The objectives you define to implement and carry out your action plan must be based on achieving more than you risk, they must be realistic, and if you cannot complete your objectives you must either change the conditions or change the objectives. Emergency workers must work within the defined plan. If they cannot work within the plan, they have an obligation to advise the person in charge why they cannot work within the plan. If the plan has to be changed, it must not be changed independently and all changes must be communicated to the incident commander in addition to all affected emergency workers. One of the most critical elements of emergency operations is that there can be only one plan in use at a time. Everyone must be made aware of the plan and what is expected of them. There can be no freelancing—freelancing creates a second (or third), independent plan. If you have been at an emergency where there were two or more plans, what did you observe? Were both the victims and the emergency responders placed in needless danger? How well were resources managed and used? With no IMS in place, operational management breaks down and people fill that void by creating their own objectives and plans to meet those objectives. If you are the incident commander, you must concentrate on managing the incident. Using the IMS structure along with your management skills in implementing the action plan will effectively allocate and protect your incident resources.

The IMS begins with the first emergency responders to arrive at the scene of an emergency. As the first trained person to arrive at an emergency, you must take control of the situation. This means taking command as you apply your plan of action. To prepare you for your role as an emergency responder, you have been trained to prepare you for your roles and responsibilities. Some of you may be specialists in what you do and may have received advanced training to prepare you. Your training gives you a confidence that lets you feel comfortable extinguishing a fire, working with a patient, or performing the other tasks you are expected to perform.

- If you are the first trained responder on the scene, are you aware that you are now the person in charge of that emergency?

    This taking charge may be limited to identifying yourself as a trained person and directing others to call 911 or make other notifications. It might also include formally establishing command if you are supervising the first arriving unit. The key is to establish leadership and direction.

- Is your first reaction to look around and assess what is going on, or are you only concentrating on the obvious?

    Size up the entire incident as quickly as possible by looking at the whole scene not just the victim or other obvious feature of the incident.

- What can you do to visibly establish control, determine your resources needs, and implement an action plan?

    When you arrive in an emergency vehicle or in uniform, you are obvious to everyone at the scene, but if it is not that obvious you will need to identify yourself. But unlike the initial identification discussed previously, you need to make sure that others arriving recognize your role. Once you are in charge, you need to assess what has happened, assess what is happening now, and then start identifying your actions options. To turn your options into an action plan, you will need resources. If you have inadequate resources, how can you get what you need?

- Do you consciously make a decision as to whether the resources you have are adequate?

    This may seem like a repeat of the previous question, but information gathering includes analysis. This is the analysis portion once you know the currently available resources.

- If they are adequate, do you have the authority to cancel other emergency responders before they arrive at the incident?

This speaks to the authority given to you as the initial incident commander. Some emergency service organizations limit the ability of lower level personnel to manage resources, which leads to the next question.

- If the emergency is beyond your ability and control, can you call for additional resources and begin planning how you will use them?

   Just because you have additional resources available doesn't mean you simply call them to the scene. It means that when you call them, you anticipate how you will be using them.

If you have been given the authority to manage resources, you have been given the authority to take control of the emergency as well as to establish command. Some emergency organizations consider the initial establishment of command as a formality until the arrival of higher ranking personnel. Establishing command with accompanying authority seriously limits what the initial commander can do.

That initial creation of the command structure and the decision-making process will set the course for the balance of the emergency. There is an old fire service axiom: "The first five minutes are more important than the next five hours." What you do in the beginning of the incident will influence everything that comes after that. Initially you may have to serve dual roles as both the IC and a rescuer, but as the incident grows, you must either pass command to someone who can serve as the IC or relinquish your role as a rescuer and become the dedicated IC.

At a large-scale emergency, the incident commander (IC) must pull back and actively manage the incident. The job of coordinating all the resources, plans, and activities is the hands-on job of the IC. As the IC there is so much you must do as the manager that if you begin performing physical tasks (setting up a tripod, treating a patient, securing the scene, etc.), you will be distracted from your role and will have relinquished command. In that case either no one will be in charge or everyone will be in charge. There can be only one incident commander. If you are the IC, determine what the needed tasks are and direct other rescuers to perform them, as shown in Figure 5-1.

FIGURE 5-1 Shown here is a group of emergency responders with one person in charge, briefing the others as to what needs to be done.

> **NOTE**
> The IC manages and directs people in order to accomplish tasks.

Regardless of your rank in your emergency organization, it is possible that you may be in charge of the scene at some point. As pointed out earlier, you may be the first person on the scene or as the incident winds down, command may be passed to you. Your responsibilities may be only that you arrive on the scene first, size up the incident, establish command, and call for help. Or as the incident winds down, you may be left in charge to get equipment back in service or secure the scene until the arrival of investigators. No matter why you were put in charge, you are now the incident manager. Do your job well, take charge of the emergency, and provide direction as to what needs to be done, who needs to do it, and when they should be doing it.

There are many reasons for using the IMS. Among these reasons are safety, unity of command, and an effective span of control.

SAFETY

Though safety should seem to be a pretty simple idea, it is not always so obvious. Your concerns will include the safety of the victim(s), emergency responders, spectators, and others who might be impacted by the emergency. You may think that addressing safety is a simple task—identify the unsafe condition and correct or control the problem. If that is how you think incident safety should be handled, you have missed a key component of safety—prevention. Prevention requires you to be proactive and recognize unsafe conditions and/or unsafe acts either before the emergency or before they occur during the emergency. In fact, NFPA 1561 requires that risk management be integrated into the regular functions of the IMS. By now you should know that confined spaces not only contain unsafe conditions but that unsafe conditions can exist outside of the space. The IC is directly responsible for incident safety, must identify the unsafe conditions, and must create a plan to control or correct unsafe acts and conditions. The IC can assign a **safety officer** to assist in managing incident safety, but the IC is ultimately responsible for managing incident safety.

> **NOTE**
> The IC must be concerned about the safety of the victim(s), emergency responders, spectators, and others who might be affected by the emergency.

UNITY OF COMMAND

Unity of command is a basic concept. There should be a single boss you are held accountable to. As simple as this seems, go back to the example of building a house given at the beginning of this chapter. Everyone wants to complete their portion of the house and get paid for their work. But what happens if you, as the general contractor, tell the carpenters to wait for the masons to finish the foundation but a carpenter supervisor tells his carpenters to begin framing when only two sides of the foundation are completed. The masons argue with the carpenters about waiting, the carpenters stop working but their supervisor comes along and tells them to get back to work and in the end, nothing gets done. No masonry foundation is completed and no framing goes up. The competing interests of each group have taken precedent. There will be competing interests at an emergency between police, fire, and EMS, and between different units within each emergency service, as shown in Figure 5-2.

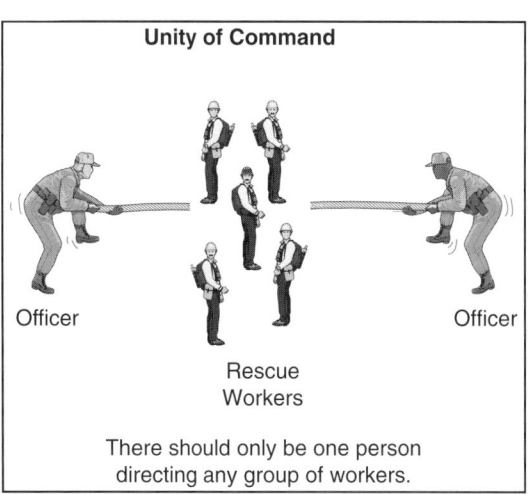

FIGURE 5-2 Unity of command is the concept that each person or group of people should have only one person directing them.

Someone has to take charge and determine what the priorities are. More importantly, you cannot allow people with incident assignments to be redirected by others who consider that their needs are more critical. When more than one person is giving orders at an emergency, it is demoralizing, counterproductive, and unsafe. A lack of unity of command can lead to conflicting, multiple plans. The IC's established goals and objectives may or may not be accomplished. If essential parts of the plan place you at risk, but the risk is controlled by actions other people are assigned to complete and those people are reassigned without the IC's knowledge, what happens to you? If a rescuer is assigned by the IC to monitor a condition such as standing next to a valve to keep it from being opened and that valve monitor is sent to get a piece of equipment a hundred feet way, who is monitoring the valve, and will the IC be aware that the valve is unmanned.

SPAN OF CONTROL

The number of people that one person can effectively manage is called the span of control. It is often expressed as a ratio of supervisors to workers. Determining the effective span of control is dependant on the task that must be completed. As the tasks become more critical or demanding, the span of control usually decreases. Where the tasks become less critical, the span of control can increase. Emergency services often use a span of control of one supervisor to between three to seven workers, as shown in Figure 5-3. How you identify the ideal span of control for an incident will be based on different factors including the supervisor's abilities (skill as a manager, experience, training, etc.), the risk involved for the work to be performed, the available communication methods, as well as the number of workers needed to complete the tasks. Routinely we see haz-mat entry teams made up of two people. This is not only a manageable span of control because of the danger of the materials involved, but it is a manageable team that can have a backup/rescue team standing by. By limiting the number of people in the hot zone, you can direct and communicate with them with little interference from others; you limit the number of people potentially exposed and ensure that you can rescue them if they get into trouble. In a confined space incident, you may have to work with a similar span of control. Depending on the circumstances in the confined space, your span of control may be as low as one supervisor to one worker or one supervisor to two workers. In a compact, tight confined space with limited accessibility, you may only be able to fit one rescuer into the space with the victim. At other times the space may be larger and you can fit more rescuers into the space, but you decide to limit the rescue team to two workers because that is the manageable number of people for that space. Regardless of how many people you commit as rescue entrants, keep the span of control manageable. You must be able to manage them safely, you must be able to communicate with them, and if the rescuers run into a problem, you must be able to rescue them.

> **NOTE**
> The greater the risk, the smaller the span of control.

One interesting note about span of control should be mentioned here. Most firefighters know about the two in, two out rule in the OSHA Respiratory Standard section (1910.134(g)(4), *Procedures for interior structural firefighting*). What you need to be aware of for confined space rescue is section 1910.134(g)(3), *Procedures for IDLH Atmospheres*. The IDLH section will affect your decisions on how many people enter the confined space to perform rescue. Under the IDLH requirements, you must have "one employee or, when needed, more than one employee . . . located outside the IDLH atmosphere." This employee must be equipped with pressure demand or other positive pressure SCBAs, or a pressure demand or other positive pressure supplied-air respirator with auxiliary SCBA; and either appropriate retrieval equipment or equivalent means for rescue. Under the IDLH section, there is no exception to allow any of the outside rescuers to be assigned other duties. In effect, the number of people you commit as rescue entrants may be limited by how much equipment you have available to provide a backup/rescue team for the rescue entrants.

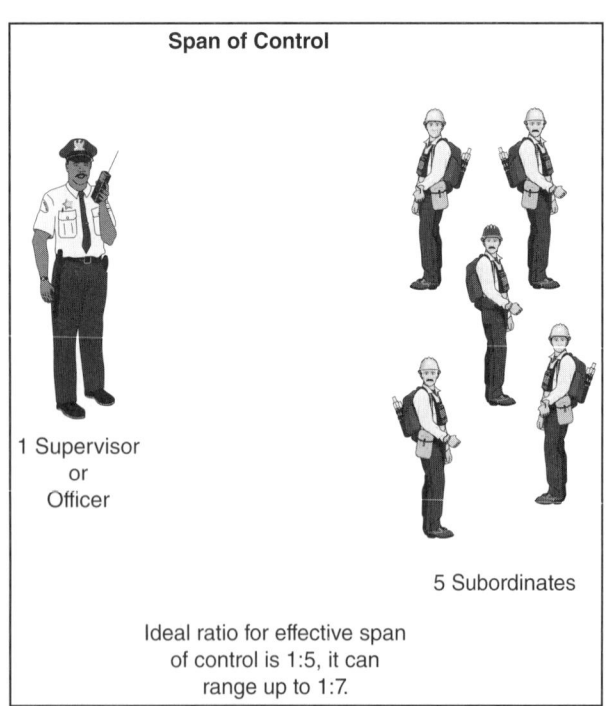

FIGURE 5-3 The span of control refers to how many people or groups of people one person can effectively supervise and direct. Ideally, the span of control is one supervisor to five workers, but depending upon the circumstances it can range from two workers to seven workers per supervisor.

COMMON TERMINOLOGY

The IMS is considered an **all risk system.** It is intended and designed to be used for all emergencies and "shall be applied to drills, exercises, and other situations that involve hazards similar to those encountered at actual emergency incidents and to simulated incidents that are conducted for training and familiarization purposes" (NFPA 1561 4.4.8). To effectively use the IMS, it must be in writing to document the system and to define roles and responsibilities. A uniform IMS should be used by all emergency service organizations (ESOs) so that each ESO can comprehend the other ESOs' roles, responsibilities, terminology, and procedures. A police officer at an incident should know who has command and, should the police officer have a fire or EMS issue that needs to be addressed, know who to bring the issue to. If a police office knows that there is an uncontrolled pipe that can carry a flammable liquid into the confined space and tells only the firefighter standing next to him about the problem, what communication has occurred? That information must get to the IC as quickly as possible and with as few filters as possible.

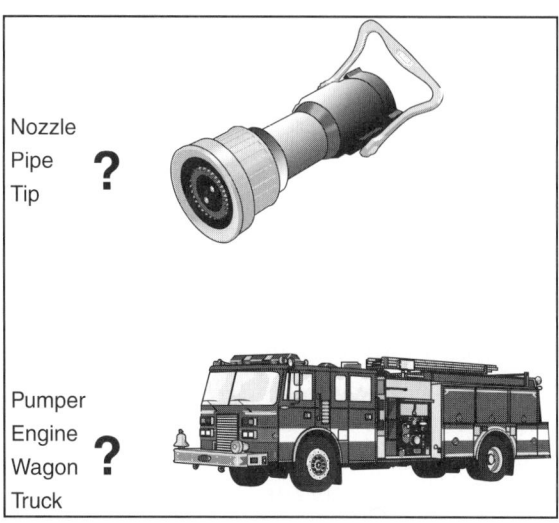

FIGURE 5-4 The same piece of equipment can have different names depending upon local preferences. Using a common terminology allows your message to be understood.

> **NOTE**
> Avoid using codes to communicate. Speak plain language so that you are understood.

There must be a vehicle for effective communication defined in the written IMS plan. Avoid using radio codes. Clear test or plain language should be used for all communications, especially between different ESOs. Make sure the person you are communicating with understands what you are telling them. There is a certain amount of jargon used within all organizations, but use caution when communicating with someone who may be from another organization. What do the terms *wagon, bus,* and *truck* mean to you? What do you think they might mean to another emergency responder? Your understanding of the term may be very different, as shown in Figure 5-4.

Successful communication requires that you send a message, that it is received, and that the person who received it understands what was said. If you have used a cell phone, you should understand the value of being heard and understood.

Operational procedures may vary between different ESOs. If you are working with multiple agencies, each agency's operational procedures can vary, even slightly, and may need to be explained during an emergency. While preplanning and joint training exercises can address these issues in advance of an incident, it doesn't always work out that way. What is important here is to make sure that others know what you will be doing and how you will do it before they assist you. Your ESO may have unique equipment that can be set up only in one particular way. You may rig your hauling system slightly differently, or other ESOs on the scene may have limited training and experience. Do not ask them to work with equipment or complete tasks they are not familiar with. The time spent establishing the methods and procedures to follow will save time and may save a rescuer's life. It is better to do it right the first time than to have to do it over again.

All ESOs have administrative tasks that must be completed for the organization to function. The people who complete these organizational tasks are the organization's managers or administrators. Managers typically function by planning, organizing, leading, and evaluating. Administrators may have weeks or months to plan out a project by setting goals and objectives. It can take days to organize the resources and then give out written assignments to complete the required goals. As the work progresses, these managers will evaluate how the plan is working and make needed adjustments. They might realize that they need to extend the time frame for completing the project or that additional resources are needed. The manager may even be able to take a vacation, delegate someone to take their place while they are gone, and come back to the project after the vacation. How does that translate to an emergency operation?

The incident commander is also a manager. She too must plan, organize, lead, and evaluate the site. Typically, though, ICs don't have hours or days to develop goals and objectives. They usually have minutes and often must work with incomplete or developing information. The life safety needs of the victims, emergency responders, and spectators require immediate attention. When ICs must develop plans based on limited and possibly inaccurate information, they must use their training and experience to guide them through the process. What has been learned from similar, past emergencies and the information about the current situation allows them to make estimates of what will happen. As the manager, the incident commander must organize information, tasks, and resources that may be changing frequently. You anticipate a need for more resources; you call for the resources, in advance, knowing it will take time to get it to the scene; and you adjust your plan to meet the schedule. At times you may not foresee the need for a particular resource, but as soon as you know you need the resource, you request it only to find out it will be delayed in arriving. You might plan on rescuers completing a task in a certain manner and find out it cannot be done the way you wanted. It might even seem as though you will never get ahead of the situation and you find that you are playing catch-up with the incident. Regardless of how you implement your plan, the incident moves on in spite of your best efforts. There will be times that might make you think that command is the most organized chaos you have ever seen.

SINGLE COMMAND AND UNIFIED COMMAND

You are called to an incident, you arrive and establish command, and your ESO is the only organization actively involved in the incident. As the IC, you have command, make the decisions, and control the action. This is an example of **single command** and it is the simplest and most common type of incident command in the IMS. One person or ESO is in charge and runs the show. That one person develops the action plan and implements the plan by assigning resources to complete tasks.

What do you do, however, when more than one ESO is actively involved in handling the incident? Will the interests of those different ESOs be allowed to compete against each other? To eliminate that competition, a **unified command** is used to develop a common set of objectives. The goal is for all the ESOs to be heard, to recognize different interests and needs, and to accommodate those needs, as shown in Figure 5-5. Working together, the members of the unified command develop an action plan and direct resources so that competition is minimized and safety and efficiency are the top priority. Examples of incidents for which unified command would be appropriate include hazardous materials incidents, large-scale emergencies, or smaller emergencies in which a crime may have been committed. The selection of the type of command to be used will depend on the IC's decision as to whether a single command or a unified command is appropriate. Regardless of which type of command is chosen, keep in mind that other ESOs consider their work as important as your ESO's work. If cooperation is critical to your success, use a unified command and consult with the other ESOs' representatives. Using a unified command does not mean that you have given up your turf or handed the emergency to someone else; it means that you have recognized the importance of other emergency response agencies in handling the emergency.

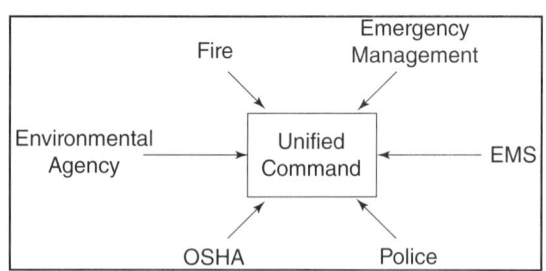

FIGURE 5-5 Unified command is the type of command that results from the contribution of many different people or agencies.

> **NOTE**
> The IC will make the decision to use single command or unified command.

THE INCIDENT ACTION PLAN

You must develop an **incident action plan (IAP)** that includes incident strategy, tactics, risk management, and safety. An IAP must be developed for every emergency to which you respond. To define your action plans, choose the goals and objectives you want to accomplish. Your goals will be fairly broad in scope and may include things such as rescue the victim. The objectives narrow the goals down to specific action items and responsibilities. Objectives identify what you would need to do and might include things such as secure the scene, set up ventilation for the victim, determine how many victims there are. To help you establish goals and objectives, you should create standard operating procedures or standard guidelines that initiate routine actions for particular types of emergencies. Most fire departments have standardized hose lays and have identified how water supplies are established. This is done so that you will not have to start from scratch at each incident. As the incident commander, SOPs mean that you can anticipate certain expected actions will have been initiated and you will have an operational basis with which you can develop the IAP. Even with established SOPs, your action plan must consider all of the strategic factors, look at the activities taking place, and determine whether they are effective. The action plan be realistic and must maintain unity of command, an effective span of control, and safety. If the IC cannot manage the situation and the people involved, then there is no action plan.

> **NOTE**
> A simple goal for a confined space accident can be to rescue the victim while protecting your personnel.

> **NOTE**
> You must have goals and objectives for your emergency operations. Your ability to manage an emergency will depend on how well you direct the emergency forces working toward the goals and objectives and how effectively they can complete them.

COMMAND POST

You will need to establish and use a **command post** at every confined space incident. What you use for a command post and how formal it needs to be will depend on the complexity of the incident. At the very least, the command post needs to be established and it must be visible. For incidents that are simple and short term, the use of a chief's car or other piece of equipment will be acceptable, but all command posts require certain basic equipment and capabilities. To be effective, your command post must have the ability to communicate with all of the radio channels used by your ESO (at more demanding incidents that communication capability must expand to all ESOs on the scene). You must identify the location of the command post and make sure that all affected parties (other ESOs, communications centers, etc.) know the location. The incident commander must remain at the command post or must designate someone to take charge in their absence. Ideally, the location of the command post should provide you with the best view of the emergency. The command post should remain stationary and not move unless conditions change significantly. You may have to abandon the position if conditions deteriorate, or you might be able to move in closer for a better command position, as shown in Figure 5-6. When the command post moves, everyone needs to know it. Larger incidents will create greater demands on the command staff and may require a formal command post to operate effectively. As the incident grows, more people are needed for IMS purposes, more ESOs and agencies will be involved, and communication needs will grow. A single radio channel or a single cell phone shared among ESOs will quickly become inadequate. The formal command post

FIGURE 5-6 When you establish a command post, identify it so that the location is obvious to everyone working at the emergency.

identifies the area where critical command activities will be taking place. It allows the IC and command staff to separate themselves from some of the distractions of the incident emergency and to focus on managing the incident. A command post that is visible also lets everyone working the incident know that someone is in charge and that the IC is accessible. Establishing a command post helps set the tone for the entire emergency—organized, managed, and professional.

RESOURCE MANAGEMENT

Resource management is one of the most critical functions for the incident commander. You have lots of "stuff" available to handle the incident. Resources include people and equipment, and you must have enough of each to sustain your action plan. Not only do you need enough stuff for what you are doing right now, you need enough to replenish it as it wears out, break down, or is consumed. The IC is responsible for budgeting the use resources, and you must spend them appropriately. Your resources already in use must be considered as invested. That is, there will be a return on your investment, a loss, or you will maintain the status quo. You certainly do not want a loss and, ideally, you do not want to maintain the status quo (except for a fully defensive operation). What you want to see is that the resources you have invested in your emergency operation are providing results. However, rarely should you plan on investing everything at once. You need to have a reserve that will allow you to reinforce your operations when needed or replace those resources as they wear out or consumed. People get tired, injured, stressed, and otherwise reach a point at which their continued deployment becomes a liability to themselves, to others around them, and to the entire operation. Continuing to use these people or any other expended resource will not improve the situation. This is when you reach for your reserves and put them to use. If you do not maintain a reserve, your operation is going to begin to suffer, and you may have to withdraw to the point that you lose all you have gained. A competent IC has a reserve to draw on, and he anticipates the maintenance needed for his resources (including people), as shown in Figure 5-7. A competent IC also realizes that providing maintenance to his now out of service resources prepares them to be employed as reserve. Once these resources have been satisfactorily recycled and recharged, the IC might be able to use them as active resources. Be aware of the status of your resources, work to stay ahead of the demands of the emergency, and invest your resources wisely. Do not run your resources (people or equipment) to failure. Doing that can have unforeseen consequences and create a domino effect through your emergency operation. Would you want to be the person on

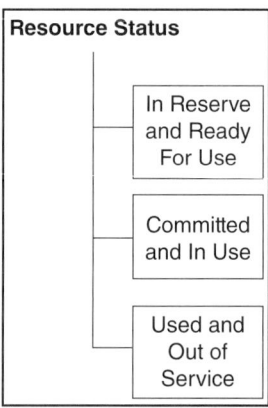

FIGURE 5-7 Establishing a staging area allows for management and coordination of resources. Any resources held in reserve will need to be put to work in an orderly fashion. That is one of the primary reasons for establishing a staging area.

the nozzle when the supply pumper runs out of fuel and the engine shuts down?

Now that you have seen the need to manage your resources, how do you manage them? The resources that initially arrive at an emergency will be directed to the scene and may very well stay right where they arrived. Typically this is what you see during fire department operations with the first alarm companies taking predetermined positions as part of the initial response procedures. But can you keep piling resources into that area? Or would it be better to keep them where you can make the most effective use of them? Having an ambulance or rescue rig buried in a group of parked vehicles delays the use of that unit. If the crew has self-dispatched elsewhere at the incident scene, you have no rig and no crew. The solution to the resource management issue is the **staging of resources**. You can handle this several ways. When you do not need to commit all of your resources to the initial alarm, you should consider staging at that point. However, if all of your first alarm resources are committed, you've called for additional resources, but have not yet decided where to deploy them; it is time to establish staging. You need to control those additional resources as they arrive. This will allow you to account for people and equipment, position them so that their presence does not interfere with emergency operations and call them up as needed.

> **NOTE**
> The three basic states in which resources exist are those that are already committed (in use), those that are in reserve for future commitment (staged), and those that have been used and are no longer available for use (expended).

Establishing a staging area keeps your reserve resources in one location, ready and available. It may be necessary to establish a forward staging area located closer to the emergency scene to reduce the lead time to deploy selected, critical resources. A good example of this type of forward staging is the use of RIT teams by the fire service. The RIT team is assigned a task (rescue of firefighters) that will be initiated only if firefighters are lost or trapped, but the team must be committed as the RIT team and staged so that they are immediately available. You can also use this for equipment you anticipate using during the emergency. When it is not possible to get apparatus close to the scene, equipment such as Sked™ stretchers, ventilation fans, spare SCBA bottles, and so on can be taken from the apparatus and brought near the scene. Whether you create one staging area or satellite staging areas for specialized resources, manage your resources. Have them ready and available so that you do not waste valuable time locating people and equipment instead of putting them to work.

INCIDENT PRIORITIES

The incident commander must constantly address and maintain three priorities. The three **incident priorities** (in order of importance) are simple:

1. Life safety
2. Incident stabilization
3. Property conservation

The IC's first priority (as it is everyone else's at the scene) must be the protection of life. That includes not only victims but also spectators and emergency responders. The IC should consider having spectators moved to a location where the IC no longer has to worry about their safety. If an emergency responder is injured, there must be a backup team (that is, a RIT or FAST team) available to rescue the injured emergency responders.

One of the most effective ways to reinforce life safety as the first priority is to require the proper use of personal protective equipment during the incident. Latex gloves, hard hats, firefighters' turnout gear, SCBA, and fall protection equipment are examples of basic personal protective equipment. The equipment is intended to prevent or minimize injuries. Make sure that it worn and set the example by using it yourself, when required.

Only after the life safety priority has been addressed can the IC concentrate on the next priority of incident stabilization. Incident stabilization is intended to keep the incident at the current size or otherwise do something to stop it from growing. To provide incident stabilization, you will have to accept that a certain amount of loss has occurred, that the loss is irreversible, and that you will begin looking for ways to control any additional loss. Whatever you do for incident stabilization, do not ignore or trade off life safety to achieve it.

Some basic examples of incident stabilization at a confined space incident include:

- Evacuation of unneeded people from the immediate area. This will free up operating space and limit the number of people potentially exposed to contaminants venting from the space.
- Initiating ventilation as soon as possible. Done properly, this can eliminate or dilute atmospheric hazards for victims within the confined space.

The third priority is property conservation. Property conservation not only means that we save property, but that we look for ways to minimize any damage that might be necessary to rescue the victims or otherwise control the emergency. It is not a priority that easily comes to mind, but it is still a priority. The simplest way to address property conservation is to do no more damage than necessary. Pre-incident planning or consultation with facility management can assist with property conservation. It may not be possible to shut down critical systems such as the drinking water supply or sewage collection and treatment without creating a significant impact on the community and causing property damage. As the third priority for the IC, you do not abandon life safety or incident stabilization to save property.

> **NOTE**
> Life safety takes priority over all other priorities, all the time. Rescue is not a blind operation but should be a planned, managed event.

These three priorities always stay in that order of importance: life safety, incident stabilization, and property conservation. Occasionally, you may be able to implement all three of them at the same time or even two of them at the same time. The order of implementation may change, but never the order of importance.

One of the most effective ways to reinforce life safety as the first priority is to require the proper use of personal protective equipment during the incident. Latex gloves, hard hats, firefighters' turnout gear, SCBA, fall protection equipment and the like are examples of basic personal protective equipment. The equipment is intended to prevent or minimize injuries. Make sure that it worn and set the example by using it yourself, when required.

COMMAND

Command is the one position that must be staffed at every emergency. Someone must be in charge, but establishing command is only the beginning of the IC's responsibilities. The IC must also set strategic goals and tactical objectives. As mentioned earlier, goals are the broad statements of what is to be accomplished. Objectives are measurable and more specific, and they identify how you will accomplish the goals. The emergency operations must have goals and objectives that are part of the action plan.

> **NOTE**
> First you must establish command.

As stated earlier in this chapter, the IC must create an action plan based on identified goals and objectives. The action plan is the road map to get you there. If you want to rescue a confined space victim and have determined that the best thing you can do at this point is to ventilate the space, the action plan should direct people to set up the fans and aim the exhaust so that it is most effective. The skills of the people performing the tasks or the critical nature of the tasks will determine how much direction is needed from the incident commander. At this point, you can see span of control creeping into the management of the incident. People with higher levels of skills can be directed with simple statements: "Take that fan and vent the space" or "Vent the space by exhausting the gases to that area." When people have lower skill levels or completing the task correctly is more critical, closer supervision, additional time to accomplish objectives, or more resources may be required. As the complexity of managing the incident grows, the IC has to consider if she can still effectively manage the incident or if she needs to assign others functional responsibilities within the IMS.

> **NOTE**
> People with higher levels of skills can be directed with objectives that are understood because of knowledge of the equipment and procedures to be used.

As the size of the emergency response team expands to meet the tactical needs of your incident, the IC must be prepared to develop a command structure to effectively manage the incident. Simple incidents require a simple command structure. Larger incidents require a larger command structure to maintain span of control and unity of command. As you expand your command structure, realize that for confined space incidents, one of the most critical areas is safety. Many potential hazards can threaten victims, emergency responders, and spectators. Consequently, you must consider, early on, if it is necessary to assign a person just to manage safety. If you assign someone a functional management role (for example, Safety Officer), everyone involved in the incident must know what the manager's role is and recognize his authority. An efficient command structure requires that incident management positions are anticipated, staffed, and recognized for their authority, as shown in Figure 5-8. That is why the IC should staff positions as needed and only as needed. Expanding the command structure requires people to staff those functional positions, and those people come from the incident resources. If you initially have only five trained responders, it is difficult to staff more than one IMS position. Call your resources early and have a plan on how you will mange the incident until the resources arrive and are in place (lead time). If you plan on stretching your resources until the needed help arrives, make sure you have your incident priorities in the proper order of importance. Once you have enough trained people on the scene to assign managerial roles, make sure they perform as expected. An IC who is inexperienced with the IMS may fail to fill the needed positions or may fill every position and not have enough people left to perform the required tasks. Keep a balance between what you have and what you need. Once you identify needed command positions, fill them as soon as possible. Reinforce that life safety takes priority over incident stabilization and make sure that everyone understands how life safety is being addressed.

The purpose of the IMS is to manage your resources so that you accomplish your mission not just efficiently, but safely. There is no efficiency without safety. Resources include people and equipment and do not overextend any resources. When the incident will be long and drawn out, anticipate the need for relief crews, food, and basic physical needs for your people. Where you have extensive equipment needs, anticipate when, where, and how you will get the required equipment and the lead time to get it to the scene and in operation where you need it. Whether your needs are for people or equipment, know the status of the resources. There are three basic states for your resources. Those that are already committed, those that are in reserve for future commitment, and those that have been used and are no longer available for use. People, just like equipment, need maintenance. Provide water, food, and rest, and address the physical needs of your people. Being well-informed about the status of your resources will let you be proactive, keep your operation moving ahead, and it might even help you to avoid Murphy's Law.

FIGURE 5-8 In this particular incident (a drill), three positions of the incident command system have been staffed: Command, Safety, and Operations.

Safety

Safety is an area of incident management that should be a continuous process. Just because the IC has assigned a safety officer, as shown in

Figure 5-9, does not mean that she has relinquished responsibility for safety. When the IC can effectively manage the incident and address incident safety without assistance, which is a good thing. The key is for the IC to recognize when assistance with safety is needed. The reality of confined space rescue is that it usually demands special attention to safety. For most emergency responders, a confined space incident is unique; it is an infrequent event, and incident conditions can present risks that are unseen without special equipment or knowledge. If there is no other reason for the IC to expand the IMS staff at a confined space incident, safety should be the one area that dictates the expansion of the IMS. The person assigned to be the safety officer should be a take-charge person.

Confined space incidents create significant and serious safety hazards to everyone present at the emergency. With the exception of the smallest confined space incident, the IC will struggle with the requirements of command and safety. Assign a safety officer, and make sure she knows she reports directly to the IC and that she is authorized to stop unsafe acts.

In addition to the previously mentioned reasons for using the IMS, coordination between the various agencies, resources, and activities is one more reason. The IC facilitates that coordination. Consider again that you are building a house—without coordination, little acceptable progress will be made by the different groups of skilled craftsmen. The IC is responsible for coordinating the use of the available resources, which may require sharing command (that is, a unified command).

Public Information Officer

At times the IC may have to deal with the media. It may be to provide information regarding incident conditions, to direct people how to respond in order to protect themselves, or simply because the story has become newsworthy. Though dealing with the press during an emergency may seem unnecessary and distracting, the press will get the information from one source or another. If you provide the information, you can ensure that it comes from a qualified source, is accurate, and is provided in a well-timed manner. If incident conditions require the public take protective actions, such as staying indoors or evacuating from selected areas, the media can assist you in getting out the proper information. At some point, the media will also want to know the identity of the victim and what caused the emergency. You want to make sure that if there are victims, their families are notified before media reports identify the victim. You also want to provide the facts concerning the incident to ensure accuracy. You, as the IC, must take responsibility for the release of public information and either you present it directly to the press or assign someone to be the **public information officer** and authorize them to release certain information. The press is part of any newsworthy event—make sure they get an accurate presentation of the facts. If nothing else, your reputation as an emergency response organization is at stake.

FIGURE 5-9 Here an individual has been assigned the role of safety officer. Safety is still the responsibility of the incident commander, but in this case, assigning a person to actively manage incident safety controls the risks associated with the incident.

NOTE

If an incident has so large an effect on the community that the public must be directed to take protective actions, the media can be a valuable asset to assist in getting out the proper information.

OTHER COMMAND SUPPORT STAFF

As the scope of the emergency grows, so do the managerial demands on the IC. At some point, during a large-scale event, the IC will have to assign other people to assist him and provide a command support staff. Many businesses start out small and grow into larger businesses. When the business grows to the point that the top managers need a staff of support personnel, they either provide the support staff or stop growing. An IC must keep up with the demands of the incident and may not be able to stop it from growing. If the IC does not provide the proper support staff, incident management begins to suffer and the entire operation may fall apart. If the IC needs help managing the incident, recognize and provide support staff. When the IC has to directly deal with issues such as safety, liaison, and public information, that is, that much less time he/she will have to devote to overall management of the emergency.

Liaison

If you have a confined space accident with serious injuries or death, there will be other governmental agencies that will need to be notified and that will respond. An OSHA or state department of labor representative may be legally required to participate in an investigation, especially if a death occurred. Medical examiners or district attorneys may become involved to determine the circumstances surrounding the event. When hazardous materials are released or there is the potential for a release to occur, the state environmental agency or department of health may take over jurisdiction at some point. The IC trying to run the emergency operation will have limited availability and may not be able to confer with every agency at an emergency scene. Assigning an individual to serve as a **liaison** to these agencies gives them access to the IC. The liaison must listen to each agency, understand its role in the incident, and then prioritize what information needs to be brought to the IC. A designated liaison officer will save the IC countless interruptions while providing the chance for each agency to be heard.

Staffing Other Functional Areas

Up to this point, we have covered the command and the command support staff. Four other functional areas are outlined in the IMS. Until someone else is assigned to manage these functional roles, the IC is directly responsible for managing those functions. The four functional areas of the IMS make up the general staff for the IC and are planning, operations, logistics, and finance/administration.

Planning

Planning collects, sorts, and interprets data about the emergency. The information can be collected from various sources and must be sorted through to determine how it might affect emergency operations. Data collection is the simple part; it is the sorting, validation, and determination of how it impacts the incident that can place special demands on this function. The IC should staff this functional area only if the demands for information gathering and interpreting exceed the IC's capability. This doesn't mean the IC lacks the ability to plan. The IC may not be able to adequately address planning because of the complexity of the incident, the location of the emergency in relationship to the command post, or some other situation that can be overcome only by assigning a planning chief.

> **NOTE**
> Planning provides incident information by collecting, sorting, and interpreting data about the emergency.

Operations

Operations takes the IC's action plan and puts it to work. Operations is task-oriented and directs the tactical components of the incident. Improving the span of control and providing effective tactical management for the resources are the most common reasons to staff the position of operations chief, as shown in Figure 5-10. If the entire emergency response consists of an incident commander and four emergency team members, there is no need to staff the operations position, and the IC should be able to effectively

FIGURE 5-10 Assigning an individual to manage the operations function of the incident is most often done when the span of control starts to become unmanageable for the incident commander.

manage this functional area. Staff operations as the incident requires.

> **NOTE**
> Operations is the functional area responsible for implementing the IC's action plan.

Logistics

Logistics provides the resources you need when and where you need them. Assigning a person to bring a meter or tripod to the entrance of the confined space is not the same as assigning a person to take charge of the logistics function. As with every other functional position in the IMS, the IC remains responsible for this area until it is assigned to someone else. When the demands for resources will exceed the ability of the incident commander to manage those resources, as shown in Figure 5-11, you should staff the logistics function. The IC must be aware that his ability to manage the incident will be affected when he is actively engaged in other activities such as moving or directing specific resources or specific talks. The IC must not compromise incident management with distractions of personally managing resources. During a small incident with few

FIGURE 5-11 This high angle rescue equipment is a good example of what can happen when no one is assigned to manage the logistics needs of an incident. There is a lot of equipment here, and it is not sorted out. How do you know if you have what you need?

resources, the IC may be effective at command and logistics. When the incident grows and the resource needs grow, the IC will have to either assign a person to be in charge of logistics or assign someone else to be in command. Don't let the functional areas suffer because you are overextended.

> **NOTE**
> Logistics is simply resource management. Getting what you need, where you need it, when you need it.

Finance/Administration

Finance/administration during an emergency is something most people do not think of as a function of the IMS. You are there to save lives and property, and who thinks about what it will cost or what paperwork needs to be filled out. At some point, the cost of the incident will have to be totaled up. If you keep track of your financial obligations and administrative needs during the emergency, it is easier than trying to recall them afterward. Knowing and recording who responded, what expendable items were used, and the name and address of any victims are finance and administration functions. As with any other functional responsibility, you staff this position only as needed.

Using the Incident Command System 73

> **NOTE**
> Most people do not think of finance/administration at an emergency as a function of the incident command system. Without it, how do you track costs for personnel equipment and supplies?

APPLYING THE INCIDENT COMMAND SYSTEM TO CONFINED SPACE RESCUE

Now that you know something about the incident management system, how do you apply it during a confined space emergency? Some people may believe that the command system needs to be large and complex, and that you should fill as many positions as possible. Though filling the command staff and general staff positions is a good idea, you must be sure that you match the demands of the incident and your available resources (people). If you have limited resources, there is no sense staffing the different positions if that leaves no one to complete the tasks. Having only five responders at the scene and assigning them the command, operations, planning, and safety positions leaves you with one person to perform all the work. If you need to staff all of those positions, you obviously anticipate a large incident, and you should be calling for more resources (people). Start by staffing command and then prioritize the staffing of the other positions. Someone has to be in charge in order to set priorities, develop an action plan and coordinate what is going to happen. By now you realize that command must be staffed, and it must be staffed at every incident.

> **NOTE**
> Staff commands at every incident.

Once you have command taken care of, determine your next need. Will it be the planning chief, safety officer, or operations chief? Look at what is the next most pressing problem that you face. Do you need tactical resources, managerial resources, or both? As the incident commander, what problems can you address and what problems do you need help with? If you need help managing the incident, assign someone to take charge of that area and assist you in managing the emergency. But what area do you address first? Take a minute to analyze the incident, the problem, and the capabilities of the emergency responders. Ask and answer the following questions:

- What is the level of knowledge and experience the rescuers have with confined space emergencies?
- Do they have the skills, but not the experience?
- Do you have to work with preconceived attitudes of the emergency responders?
- Do you have to keep them back so that they don't jump in and pull the victim out without taking the time to protect themselves?
- Or are they cautious and calculating about what they are going to do?
- Are there obvious hazards?
- What hazards are not obvious but you either know or expect are present?

> **NOTE**
> There is no sense staffing the command, operations, planning, and safety positions if you have only five emergency responders at the incident.

If you have the personnel to staff only a single position in addition to command at a confined space emergency, consider the safety officer as the next position to staff, as shown in Figure 5-12. Safety can take charge of the immediate area around the scene and isolate the area to keep out unneeded and untrained people. The safety officer can be your source for information about the atmospheric and physical hazards in and around the confined space. She can identify the need for personal protective equipment and for lockout/tagout, and she can identify and develop other ways to manage the risks at the scene. Because many of the limiting factors that will affect your operations are related to the confined space conditions, the safety officer can provide information to the incident commander to set the incident priorities and develop an action plan.

FIGURE 5-12 This is a simple Incident Command System that has been set up for a simple incident. The Command function must always be staffed regardless of the size or complexity of the incident.

> **NOTE**
> If you can staff only one other position at a confined space emergency, then you must consider making safety that position.

When you staff the other functional areas of the IMS will depend on the situation. If you get to the scene of the confined space emergency and realize that it is beyond the capabilities of your organization, you will have to call for assistance. This assistance might include a mutual aid agreement with another fire department, police department, or EMS provider for just such a contingency. These other ESOs may have training and experience far beyond yours, but that does not relieve you of command. You can pass command to them, but if you remain in command, how do you direct them? Regardless of who is in command, priorities still need to be set, goals developed, and objectives accomplished. Your organization may now be in a support role to these other ESOs, and you need to discuss what you want to accomplish with the person in charge of the other ESOs. This is a perfect situation in which unified command becomes effective. If you maintain single command, you might consider assigning the person in charge of the mutual aid group as the operations officer, as shown in Figure 5-13. When you put this person in charge of operations, it is a logical expansion of the IMS, and you have strengthened your ability to accomplish your

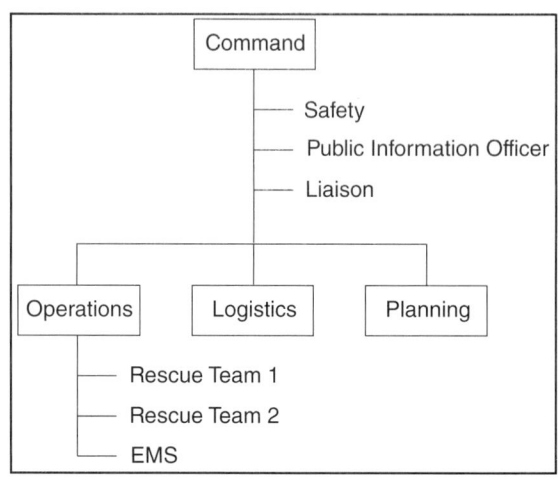

FIGURE 5-13 More complex incidents require more complex command structures to effectively handle the emergency. This structure has, hopefully, begun with Command being staffed and then expanded from there as required.

goals and objectives. Even when your organization has the training and experience to effectively handle the emergency, you may still want to staff the operation's position to maintain a manageable span of control. You want to manage the incident, not have the incident manage you. You need to be proactive, not necessarily reactive.

> **NOTE**
> The whole purpose of the incident command system is to manage the incident and not have the incident manage you.

The IC must think ahead of the emergency and anticipate how he expects the situation to evolve. If you can effectively command and get good incident information, you may not have to staff the planning function. When the IC cannot get good information and has many unanswered questions regarding the emergency, the action plan will be limited. If the quality of or the lack of information becomes a problem, get assistance. Assign responsibility to someone for the planning function. Make them the planning chief and improve your information collection and analysis. When you need to pay for additional resources or need to keep track of the costs, staff the finance/administrative function position. To effectively move and use those resources, consider the need to staff the logistics area. The key to the IMS is to grow it as your incident management needs grow. Staff only those areas when and where it will improve your ability to command the incident.

LESSONS LEARNED REVISITED

In this Lessons Learned, you have two firefighters trapped. You will now be given the opportunity to determine if they live or die. Use the following chart to identify what can happen using the IMS versus not using the IMS.

You Do Not Use The IMS

You cannot confirm if any firefighters are missing once the collapse has occurred.

Because you cannot account for personnel to determine who is missing, you spend the next 20 minutes or more trying to determine who is missing.

Because no one knows where the two firefighters were working, you must search the entire scene at the same time you fight the fire.

You find the two firefighters hours later and recover their bodies will full honors.

You Use The IMS

You can contact your companies and account for all personnel or establish who is missing.

You not only determine that 2 firefighters are missing in less than 10 minutes, you know what companies they were with and where they were working.

Because you know where the firefighters were working, you concentrate firefighting efforts in those areas to hold back the fire while your RIT team and other organized search efforts begin.

You find the firefighters alive, but injured within 30 minutes of the collapse and rescue them.

Discussion Questions

1. If you were the incident commander at the previously mentioned fire, what would happen to the action plan you implemented prior to the collapse?
2. What would be your new action plan?
3. In the expectation that many members of the fire department will want to rush to the collapse area to assist with the rescue of the trapped firefighters, how would you reorganize your incident management structure?

SUMMARY

- Someone must take charge during an emergency. Keep the control and coordination from a single source.
- Establishing command is not all there is to the IMS. The IC is responsible for safety, unity of command, the span of control, for setting the goals and objectives, and implementing the action plan.
- Expand the IMS to match the needs of the emergency.
- Know where people are, what they are doing, and if their actions are coordinated with the action plan.
- Communicate the action plan.
- Evaluate the effectiveness of the action plan.
- Use the IMS at all emergencies, regardless of size.

REVIEW QUESTIONS

1. What is the role of the IC within the incident command system? Identify the functional roles of the IMS. Which of those roles is the IC responsible for? When is it necessary to expand the IMS?
2. What is meant by an effective span of control?
3. What is unity of command?
4. Using your experience as an emergency responder and your area of specialization (police, fire, EMS, etc.), identify one way in which you routinely address the priorities of life safety, incident stabilization, and property conservation.
5. To manage an emergency operation, the IMS is designed to address which of the following?
   a. Safety
   b. Unity of command
   c. Span of control
   d. All of the above
6. The IMS is considered an all risk system. True or False?
7. The public information officer is an important function within the ICS. True or False?
8. Other functions of the ICS are
   a. Planning
   b. Operations
   c. Logistics
   d. Finance/administration
   e. All of the above
   f. Both a and b

KEY TERMS

Emergency Services Organizations (ESOs) 58
Incident Management System (IMS) 59
Incident Command System (ICS) 59
safety officer 61
all risk system 62
single command 64
unified command 64
Incident Action Plan (IAP) 65
command post 65
Resource management 66
staging of resources 67
incident priorities 67
public information officer 70
liaison 71

ACTIVITIES

Consider using a tabletop exercise that applies the IMS at a confined space incident. The focus of the exercise should be the IMS, not necessarily the confined incident. To do this, develop a scenario for the incident, create a loose script to follow, and identify benchmarks that affect the flow of the incident. The facilitator for the exercise can modify the script during the exercise as each benchmark is reached. When the benchmark is successfully completed, the facilitator simply records it and allows the incident to flow. If a benchmark has not been successfully completed on time or in sequence, the facilitator is then allowed to modify the script to match the impact of not completing a benchmark. You should also consider having people play the roles they will be expected to fill in the IMS. Because this is an IMS tabletop, fill only the management functions and staff positions. You should not have to complete any operational objectives beyond identifying them and detailing how they would be accomplished and what resources would be used.

ADDITIONAL RESOURCES

National Incident Management Courses, http://training.fema.gov/IS/NIMS.asp

National Incident Management System, http://www.fema.gov/pdf/emergency/nims/nims_doc_full.pdf

State Offices of Emergency Management

6 Strategic Rescue Factors

LESSONS LEARNED: TWO CHILDREN TRAPPED IN ABANDONED TANK

Firefighters responded to an incident in which two children were trapped in a storage tank at an abandoned manufacturing plant. The children entered a 30,000-gallon, vertical storage tank at the plastics plant that closed several years ago. The children were unable to escape from the tank, and a child who remained outside the tank went for help. Fire, police, and EMS units responded. On arrival, the rescuers found that the children had entered the tank through an 18-inch opening on the side of the tank located near the top of the 12-foot high tank. The tank contained several feet of plastic pellets.

Upon arriving at the scene, you size up the situation and identify the following critical factors:
- There are two children trapped in this confined space.
- The tank is 12 feet high and there is an 18-inch opening, the center of which is about 3 feet down from the top of the tank.
- There is another opening (manway) 30 inches in diameter, about 4 feet up from the ground, and it has a cover held in place by 12 bolts.
- Initial observations of the space show the plastic pellets are about 5 feet deep in the tank.
- You are told the children entered the tank by climbing up a makeshift ladder to the 18-inch opening, grabbed a vertical pipe inside the tank, and slid down the pipe.
- One of the children fell off the pipe and has head and leg injuries.

Critical Thinking Questions
1. What is your first priority?
2. Who will be most directly affected by this priority?
3. What is your first objective?
4. How long have the children been in the tank?

LEARNING OBJECTIVES

NFPA 1006, *2008 EDITION*
By the end of this chapter, you should be able to:

- Prepare the victim for removal from the confined space (7.1.4)
- Remove all victims using appropriate precautions and hand them over to EMS (7.1.5)
- Use a preplan to assist in size-up (7.2.1)
- Identify and evaluate the strategic limitations at a confined space incident using either a preplan of the space or information developed during incident size up (7.2.2); the preplan and/or size up information should include the following:
  - Detection and monitoring equipment, including calibration
  - Configuration of the confined space for:
    - Air sampling
    - Access for rescue
    - Victim removal
    - Continuous air monitoring
  - Identification of the contents and hazards of the space, including:
    - Materials present in the space
    - Energy sources
    - Lockout/tagout requirements
  - Other physical hazards
  - The ability to control those hazards
  - Preparations required both prior to and during entry into the confined space
  - Communication capabilities and limitations for:
    - Rescuers
    - Victims
  - Capabilities and readiness of the rescuers based on:
    - Training
    - Experience
    - Equipment
    - Psychological effects of entering and working in the space
    - Psychological and physical limitations of the PPE required for entry
    - Retrieval systems and rescue equipment
  - Accuracy and availability of incident information, including:
    - Real-time incident specific information
    - Preplan information

INTRODUCTION

On arrival at the scene of an emergency, you immediately begin to analyze the extent of the problem you are facing. In looking around, certain items can be easily identified and you catalog the effects they will have on your emergency operation. The not so obvious or more critical factors will have to be sought out, examined in more detail, and analyzed. As you conduct your assessment of the scene, you realize that some of the strategic factors can be changed, and other factors cannot be changed. How a fire building is constructed or the extent and type of injuries to a victim of a motor vehicle accident are just two examples of unchangeable factors. These unchangeable factors are called strategic factors or limiting factors, and they will define the tactics you will use to resolve the emergency. Strategic factors are so critical to your tactics and your action plan that the entire success or failure of the operation depends on identifying them, defining the impact they have, and adjusting your tactics to either work within the limits they create or finding a way to overcome them.

BASIC RESCUE SIZE-UP

A confined space rescue incident presents many strategic or limiting factors. There are so many potential strategic factors for confined spaces that an entire book could be devoted to identifying all of them. This book will not try to do that; instead, we will attempt to detail the most important ones and place them in general categories that you can use to assist in identifying the incident-specific factors.

Preplanning

Earlier in this book, you read about the importance of preplanning, and that importance cannot be overstated. Your preplan should not only identify the space and its hazards, but initial operations that you can take to secure life safety and the functional basis of an action plan. The preplan should also recognize the limits of your organization and its equipment as well as any assistance that might be needed from other agencies. We can prepare for the run of the mill events but will have to plan for large-scale or difficult rescue situations. As part of your preplanning process, you should also include the different types of confined spaces in which you may possibly work. Some of the spaces may be similar, based upon what might be common to your community (for example, silos in a farming community, septic tanks in a residential area), but other confined spaces may be one of a kind or unique to a certain facility. All confined spaces must have certain features in common to meet the definition of a confined space (large enough to enter, limited means of entry, and not designed for continuous occupancy). Start with these basics:

- Find the similarities and use that information to standardize your confined space SOPs.
- Where the confined spaces vary, look to see if that variation influences how you will operate at a rescue from that particular space.
- If the impact is significant, categorize that space so that you measure how it will affect your operations and consider how much additional planning must go into your preplan for that space.
- Where the impact is low or the space has features common to your other confined spaces, conserve your resources and limit your preplan to a common strategic plan.

Preplanning will also help you quantify your training needs. Where the spaces are similar (hazards, entry, configuration, etc.) they can be used interchangeably in preplanning, allowing you to limit what you need to maintain proficiency and keep your training efficient. When your preplanning identifies different or unique confined spaces, determine what makes them different, what common elements exist, and how your operations will be affected. Those differences will be the strategic factors that will impact your tactical operations. Based on those limiting factors, identify additional or distinct training needs, additional resource needs, and any other feature of your operational capability that will be affected. When you find these unique spaces, you should also consider if it may be possible to change conditions so that the strategic factors are mitigated.

Access for Preplanning and Training

If your ESO has been designated as the rescue team for a particular facility, there should be some sort of written agreement to that effect. Part of that agreement should include a requirement to provide access to the confined space(s) for preplanning and training. OSHA requires access for training for the designated rescue team, but cooperation between your ESO and the facility should be based on the goal of rescuing people, not simply on complying with the OSHA standards. Preparation, in advance, will not only make any potential rescue easier and faster, but it may identify requirements for pre-staging of equipment and people for a particular type of entry or remedies that reduce or eliminate hazards.

> **NOTE**
> If your agency is the designated rescue team for a particular facility, then you can require that facility to provide access to its confined space(s) for training.

BASIC STRATEGIC FACTORS

The impact of the different strategic factors at an incident will vary according to the incident, but you still need to start looking at a common set of limiting factors as you develop your action plan and begin your operations. As you consider each factor and weigh the effect of each, some factors and their tactical effects will be obvious. Some will have little effect and will be easy to dismiss, whereas others will require more scrutiny before you can measure their impact. So where do you begin your size-up?

The value of locating the confined space entry permit can not be overstated. A confined space permit, if it is present, should tell you the type of work being performed, the expected hazards, how many people may be in the confined space, who to contact for additional information (confined space supervisor or permit issuer), as well as other information that you can use during your size-up. What do you do if there is no confined space entry permit? You will have to start at ground zero, gather information, analyze it, and use what you know to develop an action plan to work your way out of the situation. The failure to provide a permit or follow basic precautions for entry into a confined space is not the fault of the rescuers, and it should not cause you to rush into a rescue. You must begin developing the information that a permit should contain; you must do it as quickly and accurately as possible; and remember the information is vital to your rescue operation, your rescuers, and the victims.

Even when a permit is present, you must check the information against your requirements.

You should consider the following strategic factors as you perform your size-up and as you conduct your emergency operations. During the course of the incident, you must evaluate the effectiveness of the action plan. If you are making progress, is it what you expected? If you are not making progress, how is it affecting your resources? If new information and/or contradictory information become available, how does that impact your action plan? The basic items you should look at initially and review periodically during the incident are:

- Atmospheric hazards
- Physical hazards
- Location and accessibility of the confined space and the victim
- Exposures
- Construction
- Contents
- Available resources for rescue operations
- Time
- Special problems
- Communications
- Life safety
- Weather conditions

Now let's look at each of these factors in more detail.

Atmospheric Hazards

Though **atmospheric hazards** seem like an obvious consideration for confined space rescue, do not take them for granted. Even after implementation of the OSHA standards designed to prevent confined space accidents, there are enough **FACE reports** and accompanying fatalities from atmospheric hazards to demonstrate that some people still haven't gotten the message. You have been called to a particular confined space emergency because something went wrong. Is the problem that one or more of the four potential atmospheric hazards exist? Are toxic gases, flammable vapors, oxygen-enriched atmospheres, or oxygen-deficient atmospheres present? When you begin your size-up, consider that any and all of the potential hazards exist in the space. Only eliminate the atmospheric hazard after you have carefully reviewed the situation by monitoring

FIGURE 6-1 Depending upon the type of work being performed within the confined space, hazards may be introduced that you might not expect.

the space and interviewing witnesses. Simply having an oxygen level within the 19.5 to 23.5 percent range or not detecting a flammable atmosphere does not eliminate all of the atmospheric hazards. You may have other gases present that are undetectable with your equipment (see Figure 6-1). Ask yourself the following questions:

- Can you determine what was occurring or what type of work was being performed in the space at the time of the accident?
- Were the entrants applying a coating and did it release toxic vapors?
- Was a tank being cleaned and as the workers reached the bottom of the tank, did they disturb a layer of sludge, thus releasing vapors?
- If part of your action plan has you venting the tank using forced ventilation, where will the exhaust gases from the tank go?
- Is it possible that ventilation will spread the atmospheric hazard?
- If you set up equipment, is it a potential source of atmospheric contaminants? A portable generator or other internal combustion–powered piece of equipment produces exhaust. Don't let the exhaust gases get carried into the confined space and create a hazardous atmosphere.

SAFETY

For an oxygen-enriched atmosphere, you want to make sure that the exhaust gases from the confined space do not come into contact with internal combustion engines, flames, or other processes **that might be an ignition source** where they can cause or accelerate combustion.

Due to potential atmospheric hazards, rescuers entering a confined space must wear positive-pressure SCBA or a **positive-pressure supplied air respirator (SAR)** with escape bottle, unless they can positively determine that no atmospheric hazard exists. SCBA or SAR must also be used when the atmospheric hazards are within permissible limits, but the potential exists for the atmosphere to become uncontrolled. The use of air purifying respirators should be avoided for rescue work.

If you are having difficulty quantifying the atmospheric hazard, take the necessary steps to protect rescuers. Wear the proper personal protective equipment (PPE) for the anticipated hazards, including chemical protective clothing if needed. The PPE becomes a strategic factor if it impacts operations, and it will. Entering a confined space through a small opening while wearing and using SCBA is difficult to do. It may create other problems for rescuers such as stability while entering through a horizontal opening. There may even be times when you will wear respiratory protection for the duration of the emergency and find out afterward that it was unnecessary. You have to go with the best information you have at the time and make the best decision that prioritizes life safety. Consider the alternative, that is, if you downplayed the atmospheric hazards and either found out conditions were more hazardous than you thought or that they worsened during entry.

NOTE

Rethink this basic philosophy: "How can I minimize the risk to operating personnel?"

It is not just the obvious hazards that present risk, but the unseen ones that you should anticipate. Atmospheric hazards are often unseen, but they are detectable if you know what to look for.

If, during your size-up of the atmospheric hazards, your detection equipment shows a flammable atmosphere or oxygen-deficient atmosphere, how do you characterize the hazard? First find out what the space contained, what the conditions were at initial entry, what periodic monitoring showed during the entry, and what work was being performed in the space. You need to identify the source of the contaminants.

Ask yourself whether the condition exist prior to the workers entering the space. Was the space purged to eliminate the condition? Did the condition change during the work entry? Or did the workers create the hazard with the work they were completing? Find out what happened to create this situation. You should also be aware that unplanned work may have been performed in the confined space or some other outside condition migrated into the space and created this problem. Determining the source of the atmospheric hazard should lead you to consider tactical options for controlling, eliminating, or working within the limits of the atmospheric hazard.

Your size-up of atmospheric hazards should help you recognize tactical options that protect the victim and the rescuers. If you are presented with an oxygen-deficient atmosphere, SCBA or SAR will probably work for the rescuers, but what can you do for the victims until the rescuers reach them? Immediate ventilation to provide fresh air to the victims may be your best tactical option, but as mentioned earlier, you must know where the exhaust gases will be going. Do not create additional hazards by directing the exhaust gases to an area where they will expand or spread the hazard. You can dilute flammable gases to get them below the lower flammable limit, but do not allow them to be vented where there is a potential ignition source. Venting gases or vapors to a low spot or an enclosed area can create a hazard to people in that area, and they may not know the contaminant is coming their way. Do you know the identity of the contaminant, and do you known its physical and chemical properties? If you are faced with an oxygen-enriched atmosphere, do not vent the exhaust gases where they can come into contact with internal combustion engines (ventilation fans, vehicle engines, etc.), flames, or other heat-producing processes. Materials that would normally be difficult to ignite or do not burn readily can catch fire and burn rapidly in an oxygen-enriched atmosphere.

Physical Hazards

When you start considering **physical hazards**, you should not limit your size-up to the need for lockout/tagout. You do need to evaluate sources of energy and equipment within the confined space. In addition, you need to weigh the hazard that equipment or energy source creates for the victim and rescuers. But don't stop at those hazards. Physical hazards exist in many forms not covered or controlled by lockout/tagout. Slippery surfaces, sharp objects, and your own equipment in the confined space can be hazards. If you use a ladder to access a confined space, secure the ladder. Falling from a ladder is a physical hazard. Are there critters inside of the confined space? Entering a confined space to rescue a victim and being met by a snake, raccoon, swarming wasp nest, or other live creature hazard must be considered. It is possible your victim entered a confined space that was safe to enter, was stung by bees, and had an allergic reaction. How do you handle that emergency? When you are looking for physical hazards, look outside of the confined space as well for potential hazards. A loose tool lying near the opening that is accidentally kicked in and hits you on the head is a physical hazard. Likewise, all of the other physical hazards mentioned previously can exist in close proximity to the confined space. Confined spaces are often located in remote areas. Imagine that your rescue team has made entry to an underground concrete vault; the victim is being brought out of the space, and the person operating the equipment at the outside edge of the vault steps into a fire ants nest. What happens to the whole operation at that point? Look for and control the physical hazards that are present, as shown in Figure 6-2.

Where the victim is located, how you can get to them, and how you can remove them from the space is the strategic factor of **location and accessibility**. This may be one of the most critical factors of your entire rescue operation. Under ideal conditions, all confined spaces would be at ground level surrounded by flat surfaces to work from, the victim would be visible from the opening, would be on a harness and retrieval line, and could be rescued by the attendant. That is the ideal, not the real world. Your victim might be in a silo with the entry point 40 or 50 feet up and on top of the silo, as shown in Figure 6-3. This is a situation where preplanning will provide a big payoff. In fact, if you are the designated rescuer team and your ESO had you review this space prior to the confined space entry, you could have either made changes to the entry or

FIGURE 6-2 This trench has a variety of physical hazards surrounding it that can easily be knocked or dropped into the space and injure people.

FIGURE 6-3 The only visible entry points to this grain elevator are at the top of the silo. This would provide a difficult rescue situation due to the location of the entry point. Pre-planning this type of rescue would save time, increase safety, and provide for a more efficient operation.

pre-staged equipment to save time. You might have even been able to suggest changes to eliminate the hazards and prevent an accident.

Regardless of what preparations were or were not made in advance, you must identify and resolve the issues you are presented with. Do you have the necessary skills and equipment to handle an emergency in a confined space with this type of access problem? If not, then you must get assistance from another ESO that does have the skills and equipment. You will also have to develop an action plan that allows you to protect the victim and rescuers (life safety priority) while stabilizing the incident until the other ESO arrives at the scene and is able to enter the confined space for rescue. Knowing your limits, knowing where to get help for more difficult incidents, and having an SOP for incident stabilization are also part of pre-planning. One of the primary goals at a confined space incident is to protect the victims and keep them alive until they can be removed from the space. We don't want victims to die because we lack the proper training or equipment to get to them and remove them, but we cannot work outside of our abilities. Poorly planned and poorly executed rescue attempts can injure or kill would-be rescuers. If you cannot locate victims or they are not accessible, you must consider how you can find them or remove them while also protecting the rescuers.

All of your rescues will not be difficult to complete. Imagine this scenario. Your rescue team was on standby duty during a confined space entry and the entrant passed out in the space. But she was on a retrieval line and could be simply hoisted out using the preset equipment. That might be all it takes to complete the rescue. You should use monitoring equipment to monitor the space and the area adjacent to it for hazards, but just like the rescue equipment, the monitoring equipment should have been preset as well. With everything already in place, you can make a rapid determination of the hazards and remove the victim without entering the space. After the victim is removed from the space, the entry permit should be revoked and conditions in the space investigated to establish what took place.

Look for simple solutions. They often work best for both you and the victim. You want to minimize how long it takes to rescue a person and minimize the risk to your people. What do you do when your victim is not visible, such as in an underground sewer line that must be entered through a manhole? Stick with the basics. Complex problems are regularly solved by putting the basics together, like building blocks. The following are among the most basic actions you can take at a confined space rescue.

1. Manage the risk to operating personnel.
2. Locate your victim. You may not even be sure you have a victim, so you will need to confirm that.
3. Identify how you will communicate with the entry team while they are in the space.
4. Determine how you will rescue the entry team, if needed.
5. Identify how you will remove the victim, if one is located.
6. Enter only with the proper PPE for the rescue entry team.

The goal is to manage the risk to everyone: rescuers, victims, and spectators. Based on your goals, create specific objectives for hazard assessment, retrieval, PPE, monitoring, medical treatment for the victim and/or rescuer, and anything else you can identify. Keep each objective as simple as possible, make sure it does not interfere with other objectives and build them together as a cohesive action plan. Make sure you properly identify the strategic factors. For example, the type of harnesses that can be used, how and where you can set up your retrieval equipment, the ability to identify atmospheric hazards, and the ability of rescuers to enter the space while wearing respiratory protection, as shown in Figure 6-4, are simple strategic factors that can be a significant impact on your rescue operation. If the rescuers must use respiratory protection as they enter the space, but your only equipment is SCBA and no one can fit through the opening with the SCBA, how do you overcome that problem? Remember, simplify your objectives to simplify your solutions. Identify who you want to do what. Provide the resources, including proper incident management.

Your size-up will require you to answer many questions. Some of the answers will be simple to find, and some will take time and hard work. As an experienced emergency worker, you probably already know that one of the most difficult and frustrating parts of an emergency is getting timely and accurate information. The information you need often comes to you in bits and pieces. It is your job to assemble the big picture. Asking the right questions and listening to and understanding the answers saves time and can mean the difference in saving lives.

FIGURE 6-4 People entering this confined space for rescue are able to enter while wearing SCBA. If the rescuers could not enter while wearing and using SCBA, how would you overcome that problem?

Exposures

In looking at a confined space emergency, can you identify potential exposures? Each agency defines exposures in a slightly different way. In general, exposures refer to nearby or adjoining areas where the incident can spread. Firefighters identify exposures as areas or buildings where the fire can spread. Hazmat technicians look at places that can be impacted through runoff, vapor clouds, or other means. For confined space rescuers, **exposures** refer to those areas where hazards, especially atmospheric hazards, from the confined space may spread. As you attempt to enter the confined space, you may ventilate the space and, as mentioned earlier, you will need to consider where these contaminants exhaust the space and where they will go once they are out of the space. As you remove a contaminated victim, do not carry out the contaminants and spread them to other places and/or people. In order to prevent spreading contaminants, consider the following:

- What can you do to contain the exposure (decontamination, directing the exhaust away from exposures, pumping product off to other approved containers, etc.).

- Consider the location of the confined space (inside or outside of a building), the elevation of the space and the potential for gases to drop or collect in low areas, the type of gases, and the characteristics of the contaminant (vapor density, flammability, etc.).
- Know the limitations of your equipment (vent hose length, volume of air it can move, power source, etc.) and whether ventilation is appropriate.

So far we've covered only atmospheric hazards. Chemical contamination of equipment is also a potential concern. You may need specialized chemical protective clothing, as shown in Figure 6-5. Prepare yourself for decontamination of the victim and/or rescuers and take stock of what decontamination supplies you have and what you might need. A successful confined space rescue in New Jersey, for example, resulted in the spread of PCBs to fire apparatus, an ambulance, and a hospital emergency room. The rescuers learned after the rescue that the space was contaminated with PCBs. Again, the timeliness and accuracy of information can be difficult, but keep working at it. Whether you suspect or have confirmed that there is a potential exposure problem, include in your action plan objectives that minimize the extent of the exposure. Include a list of likely contaminants, recognize potential routes for spreading the contaminants, and identify what will happen if the contaminants come into contact with the exposures. If having the contaminants reach an exposure is unacceptable, determine how you can prevent it.

Construction

Another basic strategic factor that should be considered is **construction**. Typically a structure is built of steel, concrete, or some other substantial material, but what if the confined space is an old brick-lined manhole leading into an underground area, or a trench that has been covered for the night, as shown in Figure 6-6? The stability of the structure of the confined space should not create additional hazards to rescuers. How would you handle an incident that was the result of an explosion of flammable gases within the space? The explosion may have caused structural damage to the space and the area surrounding it. What do you do if the opening into the space

FIGURE 6-5 Level "B" chemical protective clothing is designed to provide protection from splash hazards. The SCBA is typically worn outside of the suit, and it would be possible to wear a Class II or Class III harness with this type of protective equipment.

FIGURE 6-6 This trench is a confined space because it is covered and may contain atmospheric hazards that were trapped by the cover. However, it is still a trench, and you will still need to address any hazards that a trench presents in addition to the atmospheric hazards.

has been compromised? In these cases, the construction has become so important that your entire operation may now depend on the stability of the structure or your ability to create a new opening to gain entry. In the mid-1970s, an explosion inside of an "empty" concrete and earth-bermed, natural gas storage tank killed all of the workers inside the tank. The fire department had to remove and lift the tank roof, using heavy equipment, to retrieve the bodies. Based just on the problem of removing the tank roof, would you have the resources in personnel and equipment to handle this situation?

Contents

The **contents** of a tank, vessel, or other confined space will help you to gain some idea of the hazards you will have to face and resolve if you are called to a confined space incident. For example, an incident involving a 250,000-gallon aboveground storage tank, shown in Figure 6-7, contains water for the plant's fire protection system. This storage tank presents different hazards than a similar tank containing 250,000 gallons of acetone. Which do you think has the most hazards? Would you feel more comfortable with water or acetone? Acetone is a flammable liquid that is mildly toxic and evaporates readily. With the acetone, you need to be concerned about a flammable atmosphere, a toxic atmosphere, and a potentially oxygen-deficient atmosphere. Water, on the other hand, should be no big deal, right? Wrong! Do not ignore the obvious. Ask yourself these questions: If the tank "only" contained water, why did they call you to the scene? What work was being done inside the water tank when the accident happened? Were algaecides added to the water in the tank and either the algae or the algaecides created an atmospheric hazard? Or was the tank drained, repaired, sandblasted, and in the process of being repainted? In this case, the contents of the water tank would have changed from water to paint vapors. If you have to remove or reduce the contents of the tank to rescue or recover a victim, where can you put the contents? What do the contents weigh? Are they something heavy like sand or something light like sawdust? Will you create additional hazards by moving the contents (potential dust explosion from airborne sawdust)? Identify not only what was originally in the space, but what may have been brought into the space as a result of the work being performed.

FIGURE 6-7 This confined space can easily be taken for granted. After all, it is only a decommissioned silo slated to be cut up. What hazards can it contain when it is empty? Rescuers could easily overlook any hazards that might be present in this tank.

Resources

All of the discussion in this chapter so far has centered on confined spaces and the limitations they present. It is now time to talk about your limitations and the limitations of your equipment. Your **resources** and the capabilities of those resources will steer your operation. The number of rescuers, along with their training and experience, will determine how people can be used at a confined space incident. If their training and experience is minimal, make a good rescue entrant. Placing people in the correct role to support your operation is critical. Do not ask more of people than they are capable of doing. The capability of emergency personnel will have a direct bearing on emergency operations, as shown in Figure 6-8. If your ESO is a volunteer organization, members may work at the facility you have been called to. These members may be able to provide detailed information about the incident you are facing. For career ESOs, you may have members who worked at the facility in the past.

FIGURE 6-8 As a strategic factor, the level of training and experience of emergency personnel will have a direct bearing on emergency operations.

Most of the resource discussion thus far has been about people, but if your equipment is inadequate, you will have a hard time overcoming the deficiencies. No ESO can be prepared for every possible emergency. That is why we use mutual aid agreements. You should prepare for the types of emergencies you expect on a regular basis, but you should also plan for the rest. Know what equipment your ESO has, its designated purpose, and its proper use. You should not violate the limits set by the manufacturers or any of the various standards for different types of equipment. Overloading equipment or failing to maintain it is unacceptable. If you have identified a limitation in your equipment, understand how it will affect your operations and have a plan as to where you can get the additional resources (equipment and people trained in its proper operation) you will need. Do not simply preplan for the confined spaces you will be called to. Preplan your operation as well. Determine the resources you will need, how to obtain the resources, and what you can do, tactically, while you wait for the arrival of those resources.

Time

Waiting for the arrival of other rescuers and resources is a good example of how **time** is a strategic factor. Beyond response time, there are many other time factors to consider. Depending on the time of day, the number and types of available personnel may vary. We've already discussed response time for mutual aid, but what about the amount of time between the occurrence of the confined space accident, the discovery of the victim, and the call for help? The time of day will establish whether it is light or dark as well as changing conditions such as ambient temperature, traffic, and wind speed and direction. You can expect that the time of the year will have predictable climatic conditions. These conditions may create a need for additional personnel and resources or may influences traffic conditions. When considering time, do not neglect to include the length of time your operation will take, whether it will create additional resource needs (lighting equipment at dusk, more SCBA cylinders, fuel for equipment, etc.), and how long will it take to get to the scene and set up.

In discussing time, we must also include lead time. **Lead time** is the reflex time needed to implement and accomplish a goal or objective, as shown in Figure 6-9. Just because the incident commander has ordered something to be done does not mean that it will occur instantaneously. Orders need to be given, equipment brought to the scene and set up, and the objective achieved. It takes time for people to be told what to do, understand what is wanted, and initiate action. With lead time, coordination and timing are critical. If the objective is to ventilate a confined

FIGURE 6-9 Planning, initiating, and coordinating rescue activities take a certain amount of time known as lead time. Be sure to include this lead time when implementing your action plan.

space for 20 minutes before entry but it takes 20 minutes to get the equipment set up and running, it will be 40 minutes before the team enters. If no one is watching the time clock, your entry team might enter without any ventilation having occurred. Lead time is how long it takes to achieve the different objectives.

Communication

To have effective **communication**, the message must be sent, received, and understood. As simple as this sounds, the process is not that easy. Bring in factors such as noise, anxiety, respiratory protective equipment, and the limitations of the communication equipment, and the simplicity is lost. The interior of a confined space may create an echo that eliminates or interferes with communication. There may be physical barriers that cause the incident commander or safety officer to lose sight of the entry team. Anxiety can create a filter so that the message fits an anticipated response, or people may overreact based on a partial message. All respiratory protective equipment (SCBAs and SARs) will affect most spoken communication and may even eliminate verbal communication without special equipment. Radios, one of our most common forms of two-way communication, are limited in a variety of ways, as shown in Figure 6-10. Interference from nearby electrical equipment, the construction of the confined space, the ability of the user to transmit a clear message or transmit at all, and the strength of the battery all are possible limits on communication. When you need to identify your communication needs at a confined space emergency determines if the means of communication you have chosen let you send, receive, and understand messages between the rescue team and the incident commander or safety officer (or both). In addition, consider whether non-verbal communication is a realistic alternative.

How will the rescue team effectively communicate with other emergency personnel outside the confined space? It might mean that the rescue team members may be able to give only yes or no answers to questions. If you are using radios and you ask the rescuer a question, will they have to stop what they are doing, find and press the talk button on the radio, and then transmit the

FIGURE 6-10 At times, standard equipment used by emergency responders is difficult to use in a normal fashion. Not only must this rescuer fit through the opening, so must the SAR, flashlight, and any other needed equipment, all while suspended on the harness and ropes.

message? Doing this every time you ask a question could greatly affect their work with preparing the victim for removal from the space. Without a doubt, communication must be pre-planned. If you do not have a communication plan and practice it in advance, you run the risk of having no communication at all. As part of your communication plan, you must also have unmistakable emergency signals. Whether the emergency is inside or outside of the confined space, everyone should be able to identify and understand what the emergency signals mean.

Identifying the Risk to Life

At an emergency scene, there are three basic groups of people whose lives may be at risk, and that potential risk will need to be addressed. The three groups of people are emergency responders, victims, and spectators, as shown in Figure 6-11. You will need to identify people by the group they are part of, the extent of the hazard they face, and how you will protect them. You must protect each of these groups.

90 CHAPTER 6

FIGURE 6-11 Potential life safety hazards can occur to three groups of people: victims, rescuers, and spectators.

Emergency responders must be protected so that they are able to assist the victims. Victims must be protected as you initiate operations to rescue them so that they remain alive until rescue. Spectators must be protected so that they do not become collateral victims. Emergency responders with no active assignment who are merely watching the incident are well-trained spectators.

SAFETY
Spectators are best protected by removing them from any hazardous or potentially hazardous area. Get the spectators away so that they do not become victims if the emergency expands unexpectedly.

Weather Conditions

Weather is unique in that problems related to it can range from no problem to being one of the most critical factors. A victim trapped in a storm sewer during heavy rains may be immediately endangered by the runoff from the storm. Compare that to a victim trapped in an indoor process vessel during the same rainstorm. Though the victim may not be directly affected by the weather, your available emergency units or the response time may be affected by the weather. If you have a confined space emergency that requires you to enter from the top of an outdoor vertical storage tank and the temperature is low enough that there is a layer of frost on the top of the tank, will your team be able to safely stand or work on the surface above the entry point? Or will it be so slippery that you can not safely stand on the tank? While these may be unique weather situations, do not forget about something as simple as the temperature. Heat or cold can quickly affect your team and its performance.

NOTE
Any consideration of weather must include wind.

When you are looking at weather as a strategic factor, you must also include wind. You must determine the wind direction, as shown in Figure 6-12. Wind can carry atmospheric contaminants into, and out of, the vicinity of the confined space. If wind caries the contaminant out of the confined space, it can expose other people to the hazard. Wind speed is also important because higher wind speeds may assist by diluting the atmospheric contaminants and reduce the concentrations. At other times with low wind speed or no wind, atmospheric contaminants may be concentrated in one location and be difficult to dilute or exhaust. You must also know the direction that the wind is blowing and what factors can cause it to change direction or speed. There may be nearby geographic features, such as mountains or large bodies of water that affect wind speed or direction at different times of the day. If your operation will be of a fairly long duration and is near one of these geographic features, you must consider whether it will change conditions and affect your operation.

FIGURE 6-12 It is easy to only focus on the confined space. The flag in the upper right center of this photo is an indicator of wind direction and speed. Additionally, there is a large body of water shown in the background. What effect can this have on your operation?

Special Problems

It would be nice if we could put all the size-up considerations into a neat checklist, but just remember Murphy and his laws. This is what the category of **special problems** is related to, a kind of if it doesn't fit anywhere else, stick it in here. That makes special problems a broad category and one that you might want to dismiss. But use caution. Remember Murphy's first law: "If anything can go wrong, it will at the worst possible moment." With enough unexpected problems during an incident, you may even believe that Murphy was an optimist. Special problems range from simple problems to those that are more sophisticated. Not having electric power close to the confined space to run electrically powered lights and equipment might be a special problem. It might even be easily solved, but it doesn't fit neatly with any of the other strategic factors. Finding out that there are multiple, interconnected compartments that you must go through to get to a victim in a space is a sophisticated problem. It is possible to identify and resolve a number of special problems through preplanning, as shown in Figure 6-13, but in the real world preplanning does not always occur. When the unexpected becomes a limiting factor, you will have to evaluate the problem and determine the effect it will have on your operation.

The effect might create a ripple effect and impact other strategic factors. Regardless of how you decide to solve the problem, maintain the three priorities: life safety, incident stabilization, and property conservation. Look to break the problem into simple components and address each section individually. Collectively, these simple solutions may provide the answers for a complex operation.

Life Safety

You can address the life safety hazard to the victims by rescuing them immediately, but if you can't protect the rescuers, what have you accomplished? It is also possible to buy time for the victims as the rescuers prepare for entry (donning PPE, SCBA, etc.) by eliminating or decreasing the hazards to the victims (ventilation, etc.) and then rescuing the victims once the rescuers are protected. Ignoring the hazards to rescuers does not do anything to protect the victims. To protect the rescuers, you must do those things that decrease, eliminate, or otherwise control the risks they will face. Air monitoring in and near the confined space (as shown in Figure 6-14) helps you confirm which hazards are present and identifies the type of personal protective equipment needed to protect against

FIGURE 6-13 Pre-planning is a valuable tool for confined space rescue. These silos are interconnected by the grain elevators at the top. Grain flowing into the silo is an engulfment hazard and is a lockout/tagout issue for rescue. Pre-planning not only identifies the hazard but the methods to control the hazard.

FIGURE 6-14 You cannot control the hazards to protect rescuers and victims if you do not know the hazards within the confined space. Here, a rescuer is shown monitoring the atmosphere within the confined space.

the hazards. Managing the incident with an Incident Management System, minimizing the number of rescuers entering the confined space, and otherwise making effective use of your available resources are also examples of how you can protect rescuers. As mentioned earlier, spectators are best protected by moving them away from hazardous or potentially hazardous areas. Do not allow the spectators to become victims. If your emergency expands unexpectedly, you will not only have to worry about the original victims and exposed rescuers, but also exposed spectators. You budget your resources for the incident you have on hand. A sudden expansion of the incident can easily tax your resources. And though you should always maintain some type of resource reserve if there are spectators caught in the expansion, your reserve may not be enough for the new scenario.

> **NOTE**
> Life safety is always the first priority.

Incident Stabilization

As the number two priority, incident stabilization should not be sought at the expense of life safety. But incident stabilization can be done to benefit life safety. Ventilation is an example of how you can stabilize the incident and benefit life safety. In the previous paragraph, the initial ventilation efforts benefit the victims by diluting contaminants within the confined space, but they also provide incident stabilization because they either keep the incident from expanding (the exposure to the victims stays the same) or they reduce the incident (the victims are provided air containing a more "normal" oxygen level). Of course, the incident stabilization efforts must be realistic. Victims who are dead when you arrive at the emergency may still be dead at the conclusion of the incident whether you provide ventilation or not. Regardless of what you do for incident stabilization, keep focused on the first priority—life safety.

> **NOTE**
> The second priority is always incident stabilization.

Property Conservation

Property conservation is an ideal, a goal to work toward. It is the third priority in your incident action plan. It may not be easy to identify actions that constitute property conservation or, in a fast-paced emergency, you may not have the time to think about such an issue. No matter where property conservation ends up in your action plan, it is those actions you take to keep from doing unnecessary damage to property. The key here is what is necessary damage? Is the damage to save a life or gain incident stabilization? That damage is easy to identify as an acceptable cost. Just keep in mind that it does not mean you are free to tear apart the confined space or facility without cause. During emergency operations, firefighters routinely cause damage to property in fighting the fire, but what if the fire was in a beautiful, historic building with ornate stained glass windows? Do the firefighters break all of the windows or just what they need to break to fight the fire? There is tradeoff to be made at every incident. Do the damage you must but remember what you are trying to accomplish. Take into account the resources you have available and avoid squandering them by assigning work that achieves nothing. Both property conservation and property damage can be the result of well-intentioned hard work. Which would you prefer to spend your resources on?

> **NOTE**
> Property conservation is the third priority, which means that if it is necessary to destroy or damage property to save a life or to control the incident, then it is an acceptable cost.

Keep the incident priorities—life safety, incident stabilization, and property conservation—in the proper order of importance. Also consider the order in which you implement them. Stabilizing the incident while rescuers prepare for entry to retrieve an obviously dead victim accomplishes both life safety (for the rescuers) and incident stabilization (for the incident). Remember, though, that if you are actively working on incident stabilization and someone is injured, you must reconsider the life safety priority to make sure it is being addressed adequately or halt all activities and reevaluate what is occurring. At other incidents you will have no choice but to implement each priority in the same order as the order of importance.

LESSONS LEARNED REVISITED

Atmospheric monitoring has shown the atmosphere is acceptable for entry. Your size-up does not show any other hazards in this space, and it is not a permit-required confined space. In addition, the injured child is conscious and alert, and you have easy access to the bolts holding the 30-inch manway in place.

Discussion Questions
1. Will you need ventilation for the tank?
2. If you need to enter the tank for rescue, how will you enter?
3. What other strategic information do you need to develop an action plan for this rescue?

SUMMARY

You must identity the strategic factors of a confined space emergency before you can develop and implement an action plan. Both preplanning and the confined space entry permit are invaluable to providing accurate incident information. You must be able to identify the impact each of the following strategic factors will have on your incident:

- Atmospheric hazards
- Physical hazards
- Location and accessibility of the confined space and the victim
- Exposures
- Construction
- Contents
- Available resources for rescue operations
- Time
- Communications
- Life safety
- Weather conditions
- Special problems

Based on the strategic factors you have identified, you can then identify the incident priorities and create an action plan.

REVIEW QUESTIONS

1. Identify at least two confined spaces within your community. Describe at least four strategic factors that are present and the potential impact they will have on emergency operations. These confined spaces can be manholes, pits, silos, covered trenches, or any others in your area.
2. Based on strategic factors you identified in question 1, describe how you can either change the impact of the strategic factors or work within the limits they create.
3. If you had the opportunity to preplan the confined spaces in question 1, what information would you look for?
4. If you are the designated team for a particular facility, you can request access to its confined spaces for training. True or False?
5. In a confined space incident, you should always consider that you have an IDLH atmosphere until you determine it to be less hazardous. True or False?
6. In an oxygen-enriched atmosphere, it is not important to worry about exhaust gases coming in contact with internal combustion engines. True or False?
7. What are the basic groups of people that need to be addressed during a confined space incident?
   a. Emergency responders
   b. Victims
   c. Spectators
   d. All of the above
   e. None of the above

KEY TERMS

FACE reports 81
Positive-pressure supplied Air Respirator (SAR) 82
location and accessibility 83
physical hazards 83
exposures 85
construction 86
contents 87
resources 87
time 88
lead time 88
communications 89
weather 90
special problems 91

ACTIVITIES

1. Go to the NIOSH FACE Program home page and download several FACE reports to use as case studies to identify the strategic factors contained in each report.

2. Go to the U.S. Chemical Safety and Hazard Investigation Board's home page and download videos and reports to use for case studies.

ADDITIONAL RESOURCES

NIOSH FACE Program Web site, http://www.cdc.gov/niosh/face/default.html

The US Chemical Safety and Hazard Investigation Board video room contains both individual downloadable videos and DVD's that can be ordered free of charge, http://www.csb.gov/

7 Ventilation and Inerting

LESSONS LEARNED: TWO CHILDREN TRAPPED IN ABANDONED TANK, PART II

For this chapter, let's take the incident in Chapter 6 and expand on it. You have two children trapped in a 30,000-gallon storage tank at an abandoned manufacturing plant. The children entered the tank through an 18-inch opening on the side of the tank about 9 feet off the ground, were unable to escape, and the child with head and leg injuries is conscious and alert. The tank contained about 5 feet of plastic pellets.

You have decided to make entry through the 30-inch manway near the bottom of the tank and with some difficulty are using a reciprocating saw to cut the bolts off. The IC has ordered that continuous monitoring of the atmosphere is to be conducted near the children and the rescue attendant on a ladder is in constant contact with the victims inside the tank. As the attendant is talking to the children, the uninjured child sinks slightly into the pile of plastic pellets and tells you his feet are wet. The attendant also notices that the oxygen reading on the monitoring equipment has fallen to 20.2 percent, then 20.1 percent, and it continues to fall slowly. She immediately notifies the IC that the oxygen level is dropping.

Critical Thinking Questions
1. How have your strategic factors changed?
2. What is now your first priority?
3. What can you do to change the atmospheric conditions within the space?

LEARNING OBJECTIVES

NFPA 1006, *2008 EDITION*
By the end of this chapter, you should be able to:

- Control hazards to the rescuers by controlling hazardous materials and atmospheres and prevent further harm to the victims through the use of ventilation equipment (7.2.3)
- Describe the benefits of ventilating a confined space
- Recognize the types of equipment and methods of positive or negative pressure used for venting a confined space
- Describe the value and importance of inlet and exhaust openings, duct hoses, and other features of ventilating equipment
- Define the purpose of an inert atmosphere in a confined space

INTRODUCTION

As confined space responders, we need to be aware of the atmospheric conditions within the confined space. We may choose to ventilate the confined space to eliminate or reduce the hazards, and we should already be aware of conditions that can occur during a confined space entry to create those hazards. So far, our expectation has been that the materials, processes, or work occurring within the space have created the hazard. However, not all atmospheric hazards originate from within the space. It is possible to accidentally introduce carbon monoxide into a confined space from motor vehicle exhaust that is too close to the opening for the space. But there are also times when *intentional* changes are made to the atmospheric conditions within a confined space both before and during entry. Those intentional changes not only include ventilation, but also inerting or using another gas to displace the air within the space to intentionally eliminate oxygen. Whether ventilation is used to remove contaminants or inerting is used to remove air from the confined space, both are intentional changes to the atmosphere within the space, and the effects of each must be considered.

> **NOTE**
> Natural ventilation (unaided ventilation) of a confined space should be considered an inefficient means of venting a confined space.

VENTILATION

The **ventilation** of a confined space is intended to reduce the atmospheric contaminants by diluting or exhausting the contaminants within the space. During an emergency, ventilation can also provide victims with an air supply containing increased levels of oxygen, in addition to displacing contaminants. When you are faced with a confined space emergency, one of the first actions you want to consider is **mechanical ventilation** to protect the life of the victim. Not only does this victim-directed ventilation provide life safety and incident stabilization, it can often be accomplished quickly and defensively. By using mechanical ventilation, you do not have to enter the space or make intentional contact with the contaminants; you typically need limited PPE, and the setup of the ventilation is fairly rapid, therefore limiting the risk involved for rescuers.

As with any other action you take during an emergency, you must consider several factors before you initiate ventilation. First, you need to know the atmospheric hazards contained in the space, as shown in Figure 7-1. Is the atmosphere potentially flammable? You would not want to exhaust the gases to or through an area where ignition sources were present, just as you would not want to exhaust a toxic atmosphere to where it may come into contact with people outside of the space. Identifying potentially toxic hazards takes time, but you want to limit the exposure outside of the space. As you work to define the product and its toxicity, a simple solution may be to keep people out of areas downwind of the space or in the path of the exhaust. Depending on the incident conditions and the oxygen content of the space, you could create an oxygen-deficient or oxygen-enriched atmosphere outside the confined space. Quantifying atmospheric conditions is one of the first steps to protect both you and any victims and must be considered before initiating ventilation.

> **NOTE**
> With limited ability to ventilate a confined space, it would be better to have the air entering the space as close as possible to the victim.

> **NOTE**
> Ventilation also protects the rescuers.

Second, when you initiate ventilation operations for a confined space rescue, you must also initiate continuous monitoring of the space. Because you are attempting to change the atmospheric conditions, you must monitor to evaluate the effectiveness of your efforts. Even when your ventilation efforts are succeeding, you must continue to monitor to make sure the levels of oxygen and the reduction in contaminants are maintained at the desired concentrations.

Beyond the atmospheric condition in the space, you have several other considerations to evaluate. Look at the size of the space (as shown in Figure 7-2). What volume of air will you have to move to effectively change the atmosphere? Where will you place the fresh air inlet, and where will you locate the exhaust for air leaving the confined space? How much air can your equipment move per minute, and what are the physical properties of the contaminant?

When you start to look at the size of the confined space, you will need to estimate the volume of the space. Based on that volume, you can determine how much air you will have to move to exchange the air within the space. If the confined space is fairly small, say 1,000 cubic feet (10 feet × 10 feet × 10 feet), a mechanical blower that moves 2,500 cubic feet per minute (cfm) will change the air 2.5 times per minute. That amount of air flow will create quite a breeze within the space, but that breeze can make it difficult to work as items begin blowing around the area. Using that same 2,500-cfm fan

FIGURE 7-1 You know that this space contains a hazardous atmosphere. Now you must determine what the hazard is.

FIGURE 7-2 This confined space has a single, small (21-inch) opening for ventilation and access.

FIGURE 7-3 This large, bulk storage tank will be difficult to ventilate due to the large volume of the space.

in a round storage tank that has a volume of 1,256,000 cubic feet (a 200-foot diameter × 40-foot high tank), as shown in Figure 7-3, might be useless. With a 1,256,000 cubic feet space to ventilate, the 2,500-cfm fan will take at least 8 hours to change the atmosphere once. That doesn't mean you have to buy a fan capable of moving 1,000,000 cfm (that would be one large fan). If you have large storage tanks in your response area, it means you have to develop effective ways to use your equipment. With a large volume confined space such as a tank, you could use your 2,500-cfm fan to provide air as close as possible to the victim. This cloud of fresh air directed at the victim might be all it takes for the victim to receive the greatest benefit from the ventilation, as shown in Figure 7-4.

Inlet and Exhaust Openings

Properly locating the inlet and exhaust openings may seem fairly obvious, but you must plan the location of the exhaust for at least two reasons. First, the atmosphere in the confined space can cause a hazard to people outside as it leaves the space. Second, based on the vapor density of the gases within the confined space, the location of inlet and exhaust openings to the confined space can assist you in removing the gases from the space. Gases that are heavier than air will collect at the bottom of the space, and gases that are lighter than air will collect at the top of the space. If you place both your inlet and exhaust openings at the top of the space, how can you remove the heavier than air gases at the bottom of the space? The ideal for ventilation is to have two openings located remotely from each other, as shown in Figure 7-5. One opening could be used for the

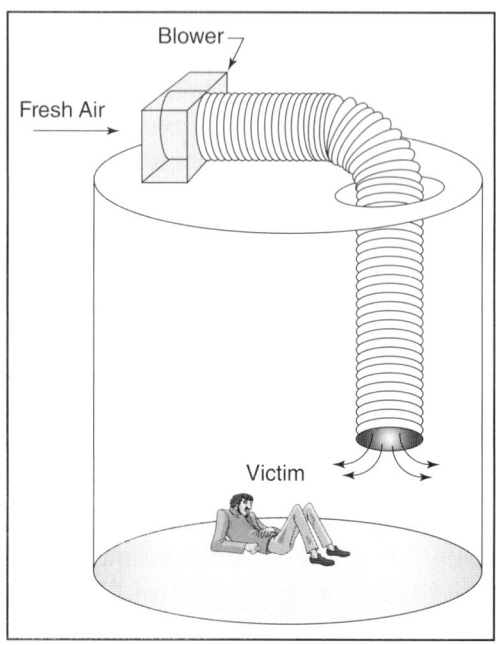

FIGURE 7-4 By blowing air into a space and placing the air hose near the victim, you can provide fresh air to the victim even if the space is large and difficult to exchange air in.

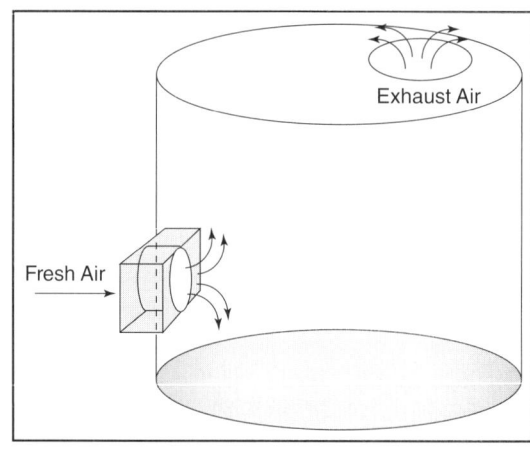

FIGURE 7-5 You must know where your exhaust gases are going as you vent a confined space. If these exhaust gases are heavier than air, or the intake opening was downwind of the exhaust opening, what effect would that have on your operation?

inlet and the other for exhaust, but, since Murphy was an optimist, we all know that scenario is often not what happens. What do you do with a confined space that has only one opening? Can you make it serve as both inlet and exhaust? The answer is yes, but with some limitations. With a single opening, only a portion of the opening can be used for the inlet. The remaining area will be your exhaust. However, this presents strategic factors that limit the effectiveness of this method. First, you need to consider churning of air, as shown in Figure 7-6. When a fan is placed in an opening and there is space around the outside of the fan, air can easily be blown out of the opening and right back through the fan. The air is essentially circulated through the fan, into the space, out of the opening, and right back through the fan. You can prevent churning by using a hose that is attached to the fan to create a remote discharge point for the air that is being moved by the fan. Ideally, this hose discharges air at a point as remote from the fan as possible, as shown in Figure 7-7. If you could place the end of this hose near the victim before rescuers enter the space, you would provide the victim with fresh air and buy time to set up and initiate the rescue operation.

Locating the exhaust opening requires other considerations. In Chapter 6, wind was identified as a strategic factor to consider during a confined space rescue. Placing the exhaust opening into the wind or placing it upwind from your inlet opening may prohibit you from effectively moving air through the space. Worse than that, wind can carry the air from the exhaust opening back to the inlet, allowing the same contaminated air to reenter the space. You must know where the exhausted gases are going after they leave the confined space. For example, when you need to vent a tank that is inside

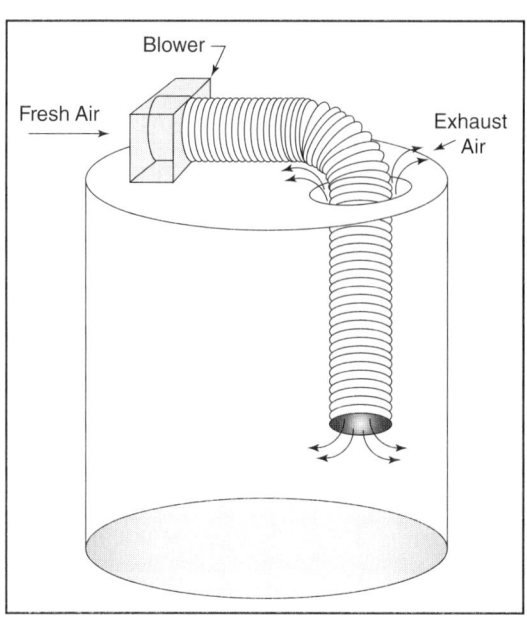

FIGURE 7-7 When you have a space with a single opening, it must be used for intake and exhaust. By using a hose to push air into the lower areas of the space shown, you will circulate air more effectively as it exhausts out of the top.

of a bermed or diked area, where will these gases go? If the exhaust gases are heavier than air, they will leave the confined space and collect in the diked area. The dike or the berm can effectively trap the exhaust gases. You have now expanded the incident area by bringing the gasses that were contained inside the confined space to the area just outside of the space. How many rescuers will be in the diked area, and how badly will they be exposed? You may find that you will have to also monitor the area outside of the confined space if exhaust gases can create a hazard in that area.

You should look to see where your victim is in relation to your available openings. Providing a flow of fresh air to the victim or diluting the contaminants within the space can buy time for your victim as you prepare to enter and make the rescue. If you respond to a confined space that contains gases that are heavier than air and your victim is lying at the bottom of the space, the victim is in the heaviest concentration of those gases. In this case, the ideal ventilation openings would be located low and near the concentrated vapors so that it is easier to remove and/or dilute the gases. Lower openings for this situation mean you will not have to lift the contaminants

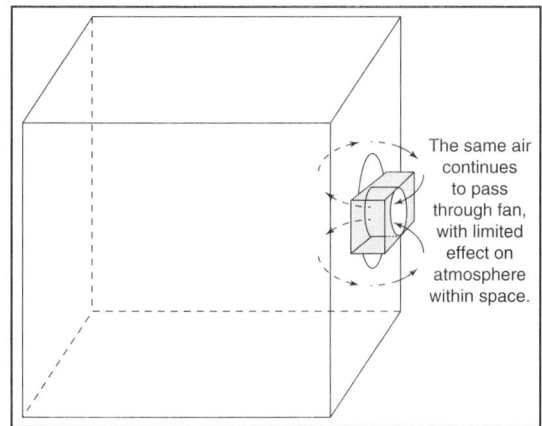

FIGURE 7-6 Churning occurs when air is blown through the fan or blower, enters and immediately exits the space, and is drawn right back through the fan. When you are venting a space, you must be aware of how effective the air movement is. Air that is churning does not contribute to ventilating the space.

out of the space and can take advantage of the fact that they are heavier than air. The opposite is true if the contaminants are lighter than air. The closer the exhaust opening is to the highest concentration of the gases in the space (based in part on the vapor density), the easier it is to vent the space. Efficient ventilation reduces the hazards not only to the victim but also to the rescuers. Look to see where your inlet and exhaust opening are located, determine how you expect the atmospheric hazards to behave (lighter or heavier than air), and take advantage of the location of your openings to vent the space.

SAFETY

Toxic atmospheres that are vented from the confined space may come into contact with people outside of the space. You must identify any potentially toxic hazards and know where the gases will migrate after they have exhausted the confined space.

What is the volume of air that your fan(s) can move? With only one fan with a capacity of 7,000 cfm, you are limited to 7,000 cfm. But what if you have three 7,000-cfm fans? How can you take advantage of that capacity? By using those three fans together, you have the potential to move 21,000 cfm of air. Depending on the site conditions, you could use those fans in tandem (one behind the other) or stack them together. Stacking the fans would be more effective than using them in tandem because each fan could reach its capacity. Fans used in tandem can lose efficiency in part due to the restrictions created by the opening the air must move through. But it is a consideration if the size or location of the inlet opening prevents you from stacking the fans. However, there are some manufacturers who have designed and built their fans to be used in tandem. You should contact the manufacturers about their specific products. When you are considering which fans or blowers to use for confined space rescue operations, you should also look at the attachments designed for the equipment. If you want to use a hose to direct air to a particular area within the space, it must be compatible with your equipment. How much hose you use will affect how much air you can move. Just like water, passing air through a hose creates turbulence that causes friction loss. The diameter and length of the air hose will affect the friction loss and the amount of air you can move. Larger-diameter and shorter hoses create less friction loss and can move more air. Smaller diameter and longer hoses create more friction loss and deliver less air. Changes in direction (such as a 90° bend) also create turbulence and will result in more friction loss.

NOTE

By placing a positive-pressure fan at the inlet and a negative-pressure fan at the exhaust, you are using two fans in tandem and will increase the airflow through the confined space.

Accessories

In 1999 when the first edition of this text was published, confined space rescue was a fairly new and developing function for many ESOs. Since then, new ventilation equipment has been developed specifically for confined spaces and new equipment will continue to come on the market. Whether your ESO will be developing a new confined space rescue team or maintaining an existing team, you must be equipped with the tools designed for confined space rescue. Equipment such as the **Saddle Vent™** (shown in Figure 7-8), heaters, and filters are available for use with ventilation equipment. The Saddle Vent™ is specifically designed for confined space use (normal entry or rescue entry), and the device looks like a flattened

FIGURE 7-8 This is a Saddle Vent®, which is designed to allow the ventilation hose to be in place and in use while providing room for people to enter the confined space. (Courtesy of Air Systems International, Inc.)

piece of air duct. Air is received from an 8-inch blower hose, passed through the Saddle Vent™ (it is meant to be placed in the opening of the confined space), and then passed into another 8-inch hose for delivery within the confined space. The Saddle Vent™ reduces the opening by only 3 inches so that entry can still be made through the primary opening.

Heaters

Heaters are available for use during winter months or in cold confined spaces. If you were faced with extreme cold, prolonged operations during cold weather, or potential hypothermia in rescuers or victims, you could heat the confined space to protect the rescuers or victims. When using heaters, you must remember that they may be potential ignition sources. When using a heater, you must follow the manufacturer's instructions and should use it in any manner that violates the listing for the equipment.

Filters

When the presence of dust can create problems during a rescue, filters are available that can be placed in line to remove the dust from either intake air or exhaust air. The use of dust filters is probably best addressed in rescue preplanning of confined spaces because you need to match the dust hazard with type of filter. Whenever dust is an issue in a confined space, you must find out if the dust is combustible. Airborne combustible dust in a confined space is a recipe for disaster, and it may take very little energy to ignite the dust and create a dust explosion.

You should also look at other accessories that are available for your ventilation equipment. Different length hoses give you flexibility in setting up the equipment, and carrying cases can help your organize and transport the equipment as a unit.

Power Sources for Ventilation Equipment

Earlier in this chapter, there was some discussion of the capabilities of ventilation equipment. That discussion was limited to the volume of air the blowers could move. Now we need to talk about two other concerns, as shown in Figure 7-9. First, what is the power source for the equipment? Blowers can be driven by a gasoline or

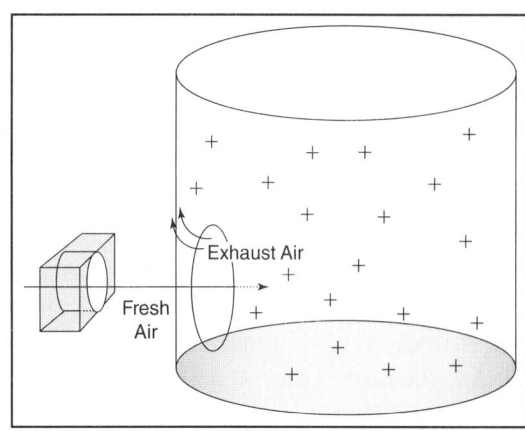

FIGURE 7-9 It is possible to use a single opening for positive-pressure ventilation. You must allow for a space at the top of the intake opening for exhaust gases to pass to the outside.

other internal combustion engine, or by an electric motor. If the driver is an internal combustion engine, you must determine if the exhaust gases will be drawn into the fan, be pushed into the confined space, and create a hazardous atmosphere. Second, can your equipment become an ignition source for flammable vapors exhausting the confined space? An internal combustion engine would seem to be an obvious ignition source when used for negative pressure ventilation, but have you considered an electrically powered fan or blower as a potential ignition source? If you are drawing flammable gases or vapors through an electric fan and the mixture is within the flammable range, the mixture can be ignited by the fan. Not only do you need to be concerned about arcing electrical components, the ignition source can be overheated equipment (bearings, bushing, belts, etc.) and static electricity (caused by the blades passing through the air). Though many electric-powered fans are intrinsically safe when you first buy them, how safe are they after they have been stored in a compartment on an emergency vehicle, used at other emergencies and contaminated with soot and carbon, or repaired by in-house mechanics? Will they still be intrinsically safe? To answer that question, talk to the manufacturer and find out what requirements exist for maintaining and testing the fan or blower.

Pneumatically powered blowers are available to be used for confined space work, but they require an air compressor for power. There are even some water-powered ventilation fans for use by the fire service. Except for unique circumstances, pneumatically powered fans and water-powered fans may be of little value at a confined space emergency. Most ESOs do not have access to the types and size compressors needed to run the pneumatic blowers. Pneumatic blowers may be available at fixed sites with on-site compressors and air supplies. Preplanning will help you to determine if they are available and if they can be of use. Due to the large amounts of water required for water-powered fans and the resulting discharge of water from the fan motor, you may find these ineffective for confined space rescue.

Potential Equipment Failures

Using powered fans or blowers to provide ventilation in support of a confined space rescue brings with it the concern the impact of an equipment failure. Remember, the primary reason for using ventilation is *life safety*. You are protecting the victims and the rescuers. If the equipment suddenly stops running, what happens? There can be simple problems, such as the electrically powered equipment becoming accidentally unplugged. Or there can be more complex problems such as loss of electrical power for that same fan. If the equipment was accidentally unplugged, it is a simple fix. Plug it back in and continue your operation. What do you do, though, if the failure was caused by a more difficult problem, such as loss of power or a broken fan motor? This is an evolving strategic factor. You need to determine if you can safely work around the problem or if you will have to pull the rescuers out of the space. Whenever there is a failure of equipment, it must be considered a new strategic factor. How you react to the problem will be based on how it affects your action plan. If the impact is significant, your entire plan may have to be scrapped or revised. If the impact is minor, make sure that it will remain a minor problem and will not have a cascading effect on your operations.

> **NOTE**
> Any failure of equipment must be considered a strategic factor and evaluated.

You should also look to minimize the impact of equipment failures by maintaining proper inspections, testing, and maintenance of your equipment. Most ESOs cannot afford to purchase and maintain a second set of ventilation equipment, so reliability becomes a critical factor. Even if you had a second set of equipment available, it would still take critical moments to get the equipment to where it was needed, set it up, and move it into operation. Something as simple as using twist lock plugs on electrically driven equipment minimizes the potential for accidentally knocking a typical three prong plug out of a socket and shutting down a blower.

Positive- and Negative-Pressure Ventilation

Which type of mechanical ventilation is better for confined space operations? Blowing air into the confined space (**positive-pressure ventilation**) or drawing air out (**negative-pressure ventilation**)? What is the difference? Let us look at each type of ventilation.

Positive-pressure ventilation takes air from outside of the space and forces it inside, as shown in Figure 7-10. By forcing the air into the space, you can use a hose to direct the airflow so as to provide fresh air near the victim. When using positive-pressure ventilation, you draw fresh air through the fan but do not contaminate it. In drawing fresh air through the fan, you also avoid the problem of potentially flammable vapors coming into contact with the fan

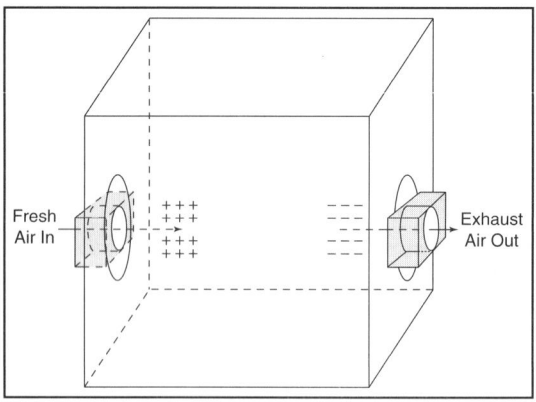

FIGURE 7-10 It is possible to use fans in combination for both positive-pressure and negative-pressure ventilation.

blades or motor. The motor of the fan is an obvious ignition source, but the spinning fan blades are less obvious. Under certain conditions, as the fan blades spin they can create a potential static electricity hazard. Under the right conditions, the static can discharge and ignite flammable vapors being drawn through the fan.

You should be aware, though, that positive-pressure ventilation is not the only type of ventilation that is useful in confined spaces. Negative-pressure ventilation can also be used with the proper precautions. You may be faced with a situation in which you will have to draw the contaminated air out of the confined space and direct it to a location remote from the inlet opening, such as when you need to avoid having the exhaust gases collect in a low lying area or work around the wind direction. If you must provide mechanical ventilation but the inlet opening is so small that it cannot effectively be used for both rescue and the placement of positive pressure ventilation equipment, you will have to look at using negative pressure ventilation through the exhaust opening. It is also possible to assist positive-pressure ventilation by using a blower at the inlet to supply air into the confined space (positive pressure) and another fan or blower to assist with exhausting air from the space (negative pressure), as shown in Figure 7-11. This creates a two-fan tandem setup and thus increases the airflow through the confined space.

FIGURE 7-11 A gasoline powered blower and two electrically powered fans. Note the size and length of the ventilation hoses as well as the different sizes of the fans.

The conditions you are presented with during an incident will determine if you use positive- or negative-pressure ventilation or some combination of both. There is no absolute statement that positive-pressure ventilation is better than negative-pressure ventilation or vice versa. There are advantages to each type of ventilation, and different people advocate one type over the other. However, keep in mind that these people are not at the incident, they are not the incident commander, and they do not have to make the choice at that exact moment. The advantages of positive-pressure ventilation often outweigh the advantages of negative-pressure ventilation, but you have to make a choice based on the real-time conditions and information you have at that exact moment during your emergency. Opt for the method that has the greatest impact on life safety, incident stabilization, and property conservation, in that order. Base your decision on how it will affect the victim's survivability and the emergency responders' safety. Don't allow the ventilation equipment to worsen the situation. If you are worried about igniting the gases exhausting form the space, do not use a method that places an ignition source in the exhaust stream. If your concern is the collection of toxic or oxygen-impacted (low or high level) exhaust gases within a diked area, move them from the area. Sometimes one method of ventilation will be the obvious choice, sometimes you will have to analyze the conditions and make the choice. Don't get locked into a tunnel vision approach. Consider all the alternatives, including combining methods if that appears to be most effective. Your choice will also depend on the capabilities of your equipment, as well as your own knowledge and experience. Ultimately, you have three choices for mechanical ventilation of a confined space: positive-pressure ventilation, negative-pressure ventilation, or a combination of the two.

> **NOTE**
> When choosing the method of ventilation for a confined space, consider the hazards of the gases, the areas or equipment they will pass through, and where they will end up once they leave the space.

Inerting

Inerting is designed to remove the oxygen from a confined space. Typical inerting gases may be gases such as nitrogen (as shown in Figure 7-12) or carbon dioxide, or a combination of gases. They are used to either eliminate one leg of the fire triangle (oxygen) to reduce the potential for fire, or they are used to stop oxidation of the product within the space and keep it from spoiling. When inerting is used, a significant amount of the inerting gas is used to replace the volume of air within the space. If you attempt to ventilate to eliminate the inerting gas, you will have to anticipate exchanging the entire atmospheric volume of the space. This could be a time-consuming process, time your victims may not have for survival.

If the atmosphere of the confined space has been inerted, you must find out why. A silo containing organic fertilizer may be subject to self-heating and spontaneous combustion. The inert atmosphere is designed to both slow down the reaction and prevent a fire. Starting up your fans and pumping fresh air with 21 percent to 22 percent oxygen into it may be all that is needed to produce flaming combustion. In that case, you will have a confined space with a fire in it. If there are victims or rescuers in the space, is that what your intention was? You must also be aware that adding air to an inerted confined space partially filled with a flammable liquid, and the ensuing vapors, may bring the vapors through the flammable range of the product. What was originally too rich to burn may now be within the flammable range.

You should be aware of one other situation in which an inert atmosphere may be created. That is the use of gaseous fire extinguishing agents within a confined space. Some spaces contain high-value equipment such as electronic switchgear or other equipment where the use of water is a concern and the primary fire suppression system uses a gaseous or clean agent system. Depending on the specific agent and the design of the system, a low oxygen atmosphere (inert) or possibly a toxic atmosphere can be produced.

Whether you discover an inerting system or fire-protection system connected to a confined space during a preplanning visit or on response to an emergency, you must immediately find out why the system is there. You must comprehend the hazard that is being protected, whether that hazard is still present, what can happen if you shut down that system, and if it can be locked out/tagged out. Depleting the oxygen with a gas such as nitrogen can cause asphyxiation. Levels below 19.5 percent oxygen are considered oxygen-deficient by OSHA. Carbon dioxide exposure at 4 percent is considered immediately dangerous to life and health by NIOSH and can be toxic at concentrations above 5 percent. Clean fire extinguishing agents are generally used in engineered systems, but the effect of exposure to these agents will vary with the agent and its properties. Some clean agents are considered toxic at levels below 15 percent. Regardless of why the inerting system or fire-protection system is part of the confined space, get control of it and keep it from becoming part of the problem. If a system must remain in service to protect against the hazard, consider removing the hazard or taking manual control of the system. If the system does not need to remain in service, get it locked out and tagged out.

You should never use inerting during a confined space rescue because you will asphyxiate the victims. The purpose of discussing inerting in this chapter is to make you aware of the potential for a confined space to contain an inert atmosphere and how it can affect your ventilation operations.

FIGURE 7-12 The white tank, next to the two silos shown in this picture, is a liquid nitrogen tank for inerting the atmosphere within the silos. Without preplanning, you may not realize that this hazard is present.

> **NOTE**
>
> Inerting is never used during a confined space rescue because the same lack of oxygen that will prevent ignition will also asphyxiate a victim.

Ventilation and Inerting

LESSONS LEARNED REVISITED

Something has occurred within the space to cause the oxygen levels to begin dropping. The levels are approaching the point at which they will be considered oxygen-deficient. If you choose to begin mechanical ventilation, you must consider several issues. The opening is 9 feet off the ground, the plastic pellets may become airborne if you aim the airstream directly at them, and the rescue opening will become the exhaust opening as soon as it is opened.

Discussion Questions

1. How will you vent the space when the only opening is 9 feet above the ground?
2. If you use a blower and hose to provide fresh air to the children, do you anticipate that the plastic pellets may become airborne?
3. The oxygen level did not begin to drop until one of the children disturbed the plastic pellets and stirred up the bottom of the tank. The child also said his feet had become wet. Could there be a relationship between stirring up the bottom of the tank and the change in oxygen readings?

SUMMARY

- Ventilation and inerting are intentional atmospheric changes within a confined space.
- Ventilation supports rescue by providing fresh air to victims and rescuers within the space.
- Ventilation can remove or dilute atmospheric contaminants within the confined space.
- The ventilation equipment must be able to do the job, and it must not contribute to the hazard.
- Quantify the levels of contaminants and oxygen so that you can define the scope of your ventilation needs.
- Continuously monitor the atmosphere within the confined space during ventilation operations.
- Control inerting systems and fire-protection systems that are connected to confined spaces you are entering.

REVIEW QUESTIONS

1. Under what circumstances would you attach a hose to your ventilation equipment and use it to direct air into a confined space?
2. What are two considerations that you must take into account when you locate your exhaust opening?
3. If you are attempting a rescue in a confined space that has a clean agent fire-protection system, would you automatically lock out and tag out the system?
4. How does "churning" of the air during ventilation affect the effectiveness of the ventilation operation?
5. During ventilation of a confined space incident, people on the outside of the space should not be concerned about the atmosphere outside the space. True or False?
6. If you place your ventilation exhaust near a running internal combustion engine, it will not create any additional hazards. True or False?
7. If you are using two blowers for ventilation at a confined space incident and one fails, it will not affect your operations. True or False?
8. You do not need to monitor the atmosphere within a confined space once your ventilation operations have been running for 30 minutes. True or False?

KEY TERMS

ventilation 96
mechanical ventilation 96
Saddle Vent™ 100

positive-pressure ventilation 102
negative-pressure ventilation 102
Inerting 104

ACTIVITIES

Activity 1
Practice using your ventilation equipment by using it to ventilate a room within a building. You can fill the room with non-toxic smoke (make sure you read the MSDS) from a smoke machine or similar equipment, you can hang lightweight strings or pieces of barrier tape in the openings, or you can inject the non-toxic smoke into the inlet of the fan when using positive pressure. Observe the movement of the smoke or the strings to see how changes in the position or location of the equipment affects air movement within the room. If the room has a window, you can open the window and use the window and a door for the inlet and exhaust openings. If you practice this on a windy day, change the position of the inlet and exhaust openings to see how the wind affects your ventilation.

Activity 2
Determine the quantity of air in cubic feet per minute (cfm) that each of your fans can move (if you have multiple fans, consider using them in tandem or positive/negative configurations). Based on the number of cfm your fans produce, determine how long it would take to completely exchange the atmosphere one time for the following spaces.

- A round tank that is 40 feet in diameter and 30 feet high
- A rectangular space that is 20 feet wide, 25 feet in length, and 12 feet high
- An oval-shaped water trailer that is 7 feet, 6 inches wide and 32 feet long

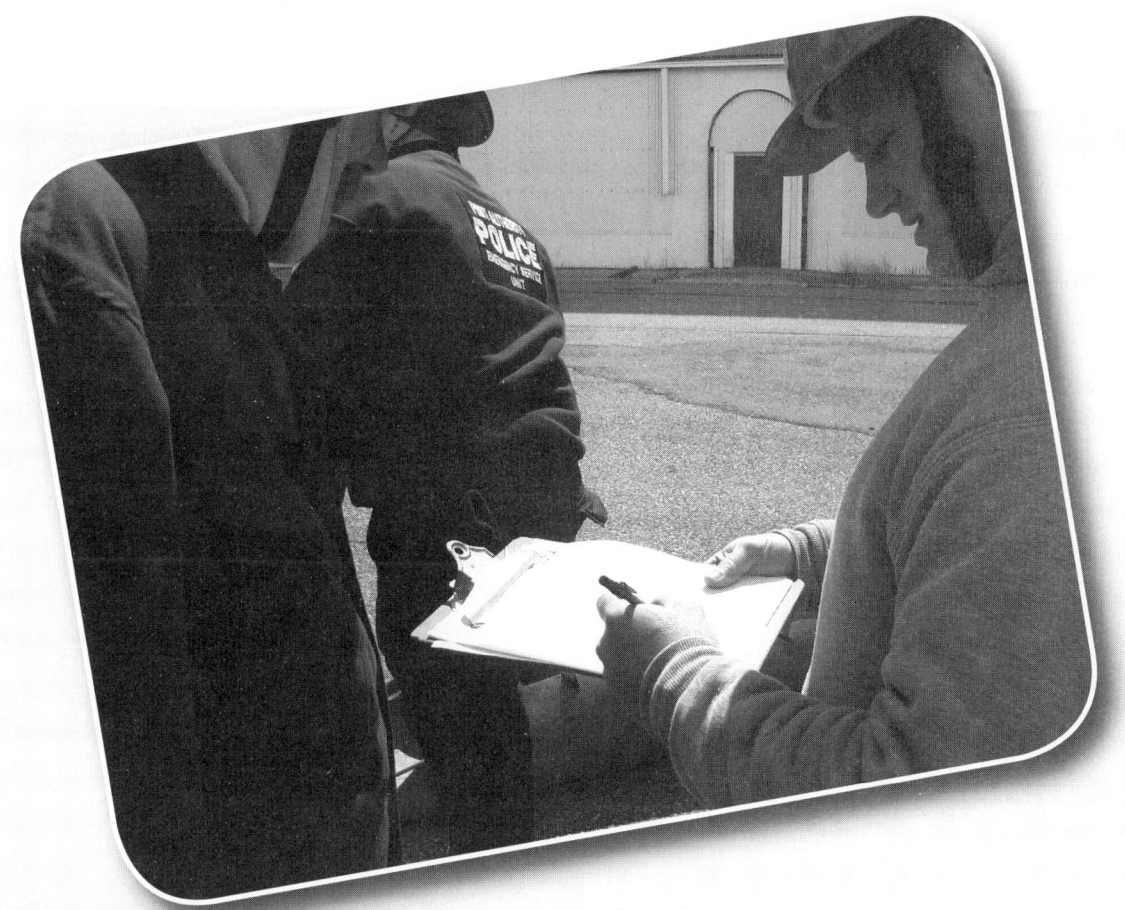

8 Safety

LESSONS LEARNED: TWO RESCUERS INJURED IN A SEWAGE PLANT TANK ACCIDENT

A firefighter and a paramedic were being treated for injuries received while rescuing a worker from an empty filtering tank at a sewage treatment plant. The worker was on a scaffold in an open-topped tank that is 15 feet wide by 24 feet long by 12 feet deep. The bottom of the tank was covered with anthracite coal used as the filtering media. Other workers in the tank stayed with the worker and called 911. Police, fire, and EMS units responded to rescue the worker. As the worker was being lifted from the tank, the firefighter collapsed. Other emergency workers in the filter tank immediately treated the firefighter and lifted her out of the tank. Shortly afterward, a paramedic reported being light-headed and nauseous but was able to climb out of the tank on her own. She received initial treatment at the scene. The weather at the time of the incident was overcast with a temperature of 90°F, high humidity, and little wind.

Critical Thinking Questions

1. How many different hazards can you identify as potential causes of the injuries to the firefighter and paramedic?
2. Of the potential causes, how would you either confirm or eliminate each?
3. Why were other workers seemingly unaffected by the conditions within the settling tank?

LEARNING OBJECTIVES

NFPA 1006, *2008 EDITION*
By the end of this chapter, you should be able to:

- Evaluate rescuer limitations (7.1.2)
- Recognize safety hazards at a confined space incident (7.1.3)

INTRODUCTION

As emergency responders, we understand that there is a certain degree of risk associated with fire, police, haz-mat, and EMS work. All too often we hear of an emergency working being injured, or worse yet, killed in the line of duty. The National Fallen Firefighters Foundation has started a safety initiative called Everyone Goes Home®. This program is based on the 16 Firefighter Life Safety Initiatives, which are designed to bring safety to the forefront and make it an integral part of the culture of the fire service. This organizational safety culture should not be restricted to firefighters. It should be part of every emergency organization. We do what we do not because of the thrill or the adrenaline rush, but because we seek to render assistance to others and make a difference in their lives. We should not be blind risk takers, but professionals who know how to manage risks. If we cannot protect ourselves, how can we protect others?

SAFETY CONSIDERATIONS FOR PERSONNEL

As you have been working your way through this book, you should now recognize that confined space rescue is not so much about the confined space as it is about the hazards unique to the space. Whole chapters have been devoted to atmospheric and physical hazards. In addition, other chapters have included OSHA requirements, roles of personnel, air monitoring, lockout/tagout, the incident command system, basic rescue size-up, and ventilation and inerting. All confined spaces must meet the three requirements to be to be classified as confined. The space must:

- Be large enough to enter and work
- Have a limited means of entry
- Not be designed for continuous employee occupancy

Permit-required confined spaces have to have only one of the following hazardous conditions to be classified as permit-required spaces:

- The potential to contain a hazardous atmosphere
- Contain a material with an engulfment potential
- Have an internal configuration that might cause entrapment or asphyxiation
- Contain any other recognized safety or health hazard

The difference between a confined space and a permit-required confined space is why when you find a hazard or hazardous condition you must identify it, as shown in Figure 8-1, analyze it to see how it impacts your operations, and control or eliminate it. That is a big challenge, and it may not be possible to identify and analyze all of the hazards present at the emergency. That doesn't mean you shouldn't look

FIGURE 8-1 Given that this is a confined space rescue scene, notice the numerous safety hazards present. These hazards include loose soil near the excavation, mechanical equipment close to the edge, no ladders for escape, and water in the space.

for hazards, it means you will have to sort them out and assign some of them precedence over others. Some of the hazards can be quickly resolved, whereas others will take so much time and effort to eliminate that you will have no chance of rescuing the victim. The use of respiratory protection during a confined space rescue in an oxygen-deficient atmosphere is one example of this. Let's say you make the decision to get to the victim quickly, but you also need to protect the rescuers; you cannot provide enough ventilation to eliminate the atmospheric hazard, so you provide the rescuers with protection. The situation is what it is, and you cannot effectively change it. You have no choice but to work within the specific limits set by the low oxygen hazard, and you can take only a few extra steps to eliminate the obstacle to the rescue. At other times, the problem can be simply fixed and the hazard eliminated, but first you must recognize it as a risk. Loose tools or equipment lying around the vertical opening to a confined space present a risk. These items can be kicked into the opening and strike people who are in the space. Eliminate or control the objects near the opening, and you will reduce the chances of them falling into the space. These are two black and white examples; unfortunately controlling the risk isn't that cut and dry. Just because you have controlled some objects that have the potential to fall into the space or provided respiratory protection doesn't mean you have eliminated the need for head protection or continuous monitoring.

If you had a confined space incident during which you were able to control a hazardous atmosphere or your monitoring showed that no atmospheric hazard existed, how comfortable would you feel, as the incident commander, about having rescuers enter without respiratory protection? There is a difference between controlled hazards and no identified hazards. If you had made the decision to ventilate a space to control an identified atmospheric hazard, but know that without the ventilation the hazardous atmosphere will reoccur, would you require respiratory protective equipment? What do you do in a situation in which there is no respiratory hazard present to begin with? The oxygen content is between 19.5 percent and 23.5 percent, there are no toxic or flammable vapors or gases to worry about (because you performed your size-up well and continuously monitored the space), and there is no dust that can become airborne. Based on your size-up, you feel pretty comfortable that you do not expect a problem to occur within the space and you determine you need only limited PPE. These are ideal conditions, but emergencies rarely produce ideal conditions. You are there because something went wrong and the conditions are not ideal for the victim. Either way, you control the conditions or the conditions control you. Your available PPE is either adequate for the conditions or it is not. If the PPE is adequate, one problem is solved. If it isn't adequate, then *why* isn't it? On a hot summer day, wearing SCBA and level B PPE (defined later in this chapter) is adequate for the respiratory and chemical exposures, but what effect will it have on the emergency services personnel wearing it? Safety is more than the obvious hazards associated with the confined space. Level B PPE worn on a hot day during a rescue will quickly drain the rescuers. It might even endanger their lives if you do not pay attention to the operational limitations it creates. The level B PPE is adequate for the hazard, but how long can your people work in it? How long can they wear it while standing around waiting for their orders or as part of a backup team? Something as simple as high ambient temperatures can double your resource needs and that one resource you need more of may be people.

By now you've read several times in this book that the first priority of the incident is life safety—that of the victims and rescuers. If you do not manage the risks to the rescuers, how can you help the victims? Managing the risks means that you must weigh the risks against the benefits to be gained: If little is to be gained, little should be risked. Emergency operations are dynamic. At times they seem to change minute by minute, and the information can be conflicting or may even contradict earlier information you received. As the incident commander, it is your job to sort through the information and scrutinize it to get to the factual basis. But to do that, you must keep a balanced approach to the risks. You cannot get tunnel vision or work on preconceived ideas of what the problem is, and you must observe everything and sort through it rapidly. Focusing only on the hazards associated with confined spaces can let something as simple as rescuers' heat injuries go unanticipated until you are suddenly confronted with them and they are now the emergency. Dramatic changes at an emergency, such as a wall collapse or injuries to rescue personnel, can cause your entire operation to immediately fall apart. You must then work to regain control of an emergency within an emergency. It is better to prevent the second emergency from occurring in the first place.

Temperature Stress

The ambient temperature conditions during rescue operations can cause injuries. If you are working outside, you must consider the air temperature as well as conditions that can affect the air temperature. Inside of the confined space, you must also be concerned with conditions that can affect the temperature within the space, as shown in Figure 8-2. Cooling or heating equipment can create high or low temperature conditions unique to the space. If you have ever gone into a walk-in freezer on a hot summer day, it doesn't seem too bad at first, but it quickly becomes too cold to stay in for very

FIGURE 8-2 Confined spaces can be above or below the ambient temperature. Shown here is a steam coil for heating the product normally stored in this tank.

long. The temperature can be more than 100°F outside the freezer, but once you are in the freezer it may as well be winter. No matter what causes the air temperature to go up or down, you must think about what it will do to the rescue personnel and how it will affect your rescue operation. Exposure to high or low temperatures can affect people's ability to aid in the rescue. Rescuers can also become victims in need of medical treatment or rescue. Whether people are inside or outside of the confined space, they can be affected by temperature. It is a matter of how long they are exposed, the severity of the temperature conditions, and how prepared they are to deal with the conditions.

NOTE
High or low air temperatures can injure rescuers.

Humans can work efficiently at ambient temperatures up to 78°F. Once the temperature goes above 78°F, our efficiency begins to decrease from the effects of heat. Add in other factors such as age, physical condition, sex, and protective equipment being worn, and the effects of heat are amplified. We often tend to think of heat becoming a problem when the temperature rises extraordinarily and creates a "heat wave." That is not always the case. Air temperatures as low as 80°F combined with high humidity can make the air temperature feel the equivalent of 90°F and above. This is known as the apparent temperature, and the body reacts as though the

FIGURE 8-3 Minimize the number of people who are needed to perform the rescue safely. Here is a small elevated platform crowded with rescuers.

temperature really was 90°F or above. Include this with the other factors listed above (age, physical condition, impervious clothing, etc.), and you'll see that heat stress injuries can occur before the potential hazard is recognized. Elevated temperatures expose rescue personnel to the risk of heat cramps, heat exhaustion, or heat stroke. Preventing heat-related injuries is the best method of dealing with heat stress. To avoid heat stress, you must take steps to reduce the impact that heat has on the human body. These preventive steps include:

- Changing the ambient conditions within the confined space by heating or cooling the atmosphere within the space
- Limiting the number of people exposed to the heat (including direct sunlight) by keeping people out of the heat, as shown in Figure 8-3
- Using fans to blow air across people to cool them
- Having entry team members drink liquids both before and after entry into the confined space to replace lost fluids
- Managing the amount of time that personnel must wear impervious protective clothing, as shown in Figure 8-4
- Implementing an action plan that has people pace their work to match conditions
- Medical monitoring of rescue personnel as conditions dictate

FIGURE 8-4 Protective clothing can increase the potential for heat stress injuries to rescuers. These Level B suits can trap moisture from sweat and slow down the evaporative cooling effect.

Mechanical ventilation can provide cooling. Blowing fresh, clean, cooler air into the confined space assists the victim and the rescuers, and it can have an ancillary benefit because it also cools people within the space. Just remember that if you are ventilating to cool the space, you still must take precautions needed for any ventilation of a confined space. Know the source of your supply air, know where the air will exhaust, and evaluate the effects within the space. Determine whether you can, or want to, exchange the air within the space. In summer months, both the inside and outside temperatures of a confined space may be high. In that case, all you might be doing is moving hot air. If you have a tank with limited entry and ventilation openings, effective air movement for cooling may be difficult. Similarly, it may be difficult to cool a large tank. In such cases, you will need to be resourceful. Directing the cooling air so that it is blowing across the rescuers within the space may be the most efficient way to provide cooling. Keep in mind the safety of the victim and the condition of the atmosphere within the space. If you are blowing so much air into the space that it creates airborne debris and dust, you may create more problems than you solve. An even simpler limitation may be that you can use only one fan, and its use is limited to providing ventilation and fresh air to the victim.

Keeping people out of the heat sounds so simple that it almost doesn't need to be mentioned. Despite that simplicity, support personnel (the incident commander, standby personnel, and support people working away from the incident site) can become victims of heat stress. Manage where you set up the command post or where you stage people. Staying out of the heat and out of direct sunlight should be a concern in hot weather. If you have to, set up some type of shade covering. If it is difficult to provide shade or the temperatures are very high, use a fan to provide air movement. We sweat to cool our bodies. Sweat provides evaporative cooling. Air movement increases evaporation, which is why you feel so much cooler in a breeze when you are sweating. Backup personnel waiting in the sun for their chance to work are being exposed to the heat. This preheating reduces the length of time they can function effectively when they are called to work. A balanced approach to managing the risk of heat injuries includes staying out of the heat to maximize your available work time.

> **NOTE**
>
> Minimizing the number of people exposed to the heat sounds too simple to mention. Despite that simplicity, it is not unheard of for support personnel (the incident commander, standby personnel, support people working away from the incident site) to become victims of heat stress.

Stay Hydrated

Hydrate, hydrate, hydrate. Sweating reduces the volume of liquid within the body. You must do something to replace those liquids, as shown in Figure 8-5. Reduce the volume of liquids in the body too much, and the body will take steps to protect itself. Heat cramps, heat exhaustion, and heat stroke can follow. To prevent heat injuries, begin with the basics. Have entry team members drink water or a sports drink to replace electrolytes, both before and after entry. Base the amount of liquid on the conditions you are faced with. You must consider not only the actual

FIGURE 8-5 As you sweat, you lose fluids. You must replace those fluids to help prevent heat-related injuries.

temperature, but also the apparent temperature to avoid heat injuries. When hydrating, be aware that it can take 20 minutes or more for the liquid to be available for the body to use. Drinking liquids before entry prepares you so that you will have the liquids available while you are working. Drinking liquids after leaving the confined space replaces lost liquids. Have the entry team (and anyone else working under similar conditions) hydrate even if they state that they are not thirsty. Thirst may not be a reliable indicator of dehydration, and if you wait until you are thirsty, you may be well on your way to heat stress.

Dress Appropriately

Most emergency workers wear some type of protective clothing. Bulletproof vests, turnout gear, and other types of personal protective equipment are designed for a specific hazard or hazards. While fighting a fire, firefighters need to protect their skin from heat and steam. Unfortunately, as good as that equipment is at protecting you from the external heat and moisture, it is equally as good at holding in body heat and preventing sweat from evaporating. Other emergency workers may be wearing clothing that is designed to keep out the rain or cold and it can be impervious, just like firefighters' protective clothing. Though this equipment may be good at protecting you from the "normal" hazards of your job, what problems can it create at an "abnormal" emergency such as a confined space rescue? Yes, you will have to wear some type of protective equipment, but is it the right equipment for the hazards you will be facing? Dress appropriately for the incident conditions. Wear what protective clothing is needed, but make sure you consider the effect that the clothing will have on personnel. High temperatures and impervious clothing create difficult tactical conditions for emergency workers. Their efficiency will decrease, it will take longer to accomplish the assigned tasks, and it will most likely take more people to accomplish those tasks. Consider if it is appropriate to have rescue personnel who are staged outside of the hazard area open up or remove the protective clothing while they are awaiting assignments. People assigned as the backup or rescue team to the entry team will have to continue to wear their protective clothing so that they can be quickly deployed. Keep them out of the heat as much as possible and dress them only to the point at which they can quickly finish donning the protective equipment and immediately go to work.

Pace Yourself

Pacing the work of the rescue team and having realistic expectations as to what they can accomplish is part of safety. Matching the work to the conditions is always a good idea, regardless of the temperature. As the incident commander, assign work to people that they can accomplish based on the working conditions. And enforce those decisions. Do not let people overextend themselves. That does not mean you discourage efficiency. It means that you monitor people to keep them from taking shortcuts or rushing so that they do not injure themselves or others. Remember the discussion in the Incident Management System section about span of control? This is an example of where the span of control may have to be reduced for the sake of safety. The goal of the action plan is to accomplish the needed objectives, prevent injuries, and provide for a flow to the emergency operation so that you maintain the required level of control, and thus succeed.

> **NOTE**
> Having rescue workers pace their work to match conditions at an emergency is always a good idea regardless of the temperature.

Medical Monitoring

Medical monitoring of rescue personnel is not just an intervention, but prevention. It may provide the first indicator of a developing medical emergency, or it may identify an emergency worker who should not be assigned certain tasks. Medical monitoring does not replace managing the risks at an incident, it only helps you to do that. In addition to having EMS on the scene for the victims, you should have EMS personnel on the scene who are assigned to stand by to treat rescue personnel. These EMS resources are dedicated to the rescue workers.

Heat-Related Injuries

Will you have incident-specific hazards that might be best managed through medical monitoring? Yes, you will, and heat stress is one of those hazards that requires medical monitoring both before entry into the confined space and after exiting the space. Comparing pre-entry and post-exit vital signs such as pulse, blood pressure, respirations, and body temperature can indicate the onset of heat injuries. Conducting medical monitoring reduces the risk of heat injuries by giving real-time information as to what is going on within the body. An increase in body temperature between the pre-entry and the post-exit monitoring may indicate the need for follow-up monitoring or immediate medical attention to prevent further heat injuries. Prevention of a heat stress injury by hydration and monitoring is the key factor. Left untreated, heat stroke can cause death.

Though prevention of heat stress injuries is important, you must also be able to recognize the signs and symptoms of heat stress, shown in Figure 8-6. Knowing the signs and symptoms of heat stress lets you evaluate your own physical condition as well as the condition of others around you. The symptoms of heat cramps include severe muscle cramps in the abdomen and legs, and they are believed to be caused by loss of fluids from sweating heavily. Heat cramps may be the first sign of heat stress, but, depending on how severe the heat injury is, they may not. If your heat injury is severe enough, you may go right past heat cramps and exhibit symptoms of heat exhaustion or heat stroke. Victims of heat exhaustion may feel weak, nauseous, dizzy, faint, and have pale and clammy skin as the blood

> **Signs and Symptoms of Heat-Related Injuries**
>
> *Heat Cramps*
>
> Muscle cramps—especially legs and abdomen
>
> Normal body temperature and moist skin
>
> Can advance to other more serious heat-related injuries
>
> *Heat Exhaustion*
>
> Headache, nausea, dizziness
>
> Exhaustion
>
> Normal or below normal body temperature and cool, moist skin
>
> Can advance to heat stroke
>
> *Heat Stroke*
>
> May be unconscious
>
> Hot, dry, red skin
>
> High body temperature
>
> Can lead to convulsions, coma, and death

FIGURE 8-6 When working in conditions that can cause heat-related injuries, you must recognize the signs and symptoms of heat cramps, heat exhaustion, and heat stroke.

flows to the skin and away from vital organs in an attempt to cool the body. Heavy perspiration and normal or below normal body temperature also occur. Heat exhaustion victims rarely lose consciousness, but they may. If not treated, heat exhaustion can progress to heat stroke. Heat stroke is a true medical emergency and is the most serious heat injury. Without immediate medical attention, people can suffer serious injury to vital organs, and up to 20 percent of heat stroke victims die from it. Heat stroke signs and symptoms include lack of sweating (even though the skin may be wet from earlier perspiration); dry, hot, red skin; elevated body temperature; and possible convulsions or unconsciousness. The body can no longer regulate temperature, and the body temperature may rise to 104°F or more.

Cold-Related Injuries

Up to now, we have been addressing weather-related injuries involving heat. You must also be concerned about injuries from cold temperatures. Acutely, cold injuries may be less serious than heat injuries, but just as for heat injuries, you must consider the hazard presented by cold temperatures and work to prevent injury. Cold conditions may be present because of the weather or from intentional or accidental conditions in the confined space, such as a refrigerated hold on a ship. As the ambient temperature drops, the body attempts to produce heat to keep up with the loss of heat to the cold. When the body cannot produce enough heat to stay warm, the blood supply retreats to the head and torso in an attempt to protect the brain and vital organs. This in turn leads to shivering, and you begin to lose muscle coordination. When the condition lasts long enough that the body core temperature begins to drop, it leads to **hypothermia**. If the hypothermia is severe enough, unconsciousness and death can follow. It does not have to be extremely cold for hypothermia to occur. A person wearing wet clothing in air with a temperature of 50°F can suffer hypothermia.

Protection against cold temperatures and cold-induced injuries includes dressing for the temperature to prevent loss of body heat, not allowing your clothing to get wet from your own sweat or the liquids in the confined space, and pacing your work. Recognizing a cold temperature hazard may be difficult during warmer weather. As pointed out earlier, under the right conditions, 50°F is low enough to cause hypothermia and on a warm winter day 50° feels pretty warm. If you have preplanned the confined space and know a cold temperature hazard is likely to be present, you can take steps, in advance, to address the problem. If you have not had the chance to preplan, you may only discover the cold hazard during your size-up. As with heat, cold affects tactical operations. Extra layers of clothing and gloves limit mobility and dexterity. You must evaluate the cold to determine how much of a strategic factor it produces. Just as with high temperatures, prevention of the injury should be your goal.

We discussed earlier that to prevent heat injuries, fans or blowers can be used to cool people. Now we have the reverse problem. We do not want to cool people, we want to keep them from getting colder and possibly even warm them. This being said, air movement, especially

Windchill Factor								
Actual Temp (F degrees)				Wind (miles per hour)				
Calm	5	10	15	20	25	30	35	40
50	48	40	36	32	30	28	27	26
40	37	28	22	18	16	13	11	10
30	27	16	9	4	0	−2	−4	−6
20	16	4	−5	−10	−15	−18	−20	−21
10	6	−9	−18	−25	−29	−33	−35	−37
0	−5	−21	−36	−39	−44	−48	−49	−53
−10	−15	−33	−45	−53	−59	−63	−67	−69
−20	−26	−46	−58	−67	−74	−79	−82	−85
−30	−36	−58	−72	−82	−87	−94	−98	−102

FIGURE 8-7 Wind chill chart. Notice how the wind can create a cooling effect at mild temperatures.

the wind, now becomes your enemy. Most of us have heard the term *wind-chill,* and if you live in a colder climate, wind-chill information is generally included in winter weather reports. As air circulates around the body, it draws off heat, as shown in Figure 8-7. Typically when we think of wind, we think of the naturally occurring wind in the environment. To recognize wind-chill and how it can add to cold injury potential, you must think of any source of air movement. Moving air from blowers used for mechanical ventilation can produce a wind-chill effect as it blows across you. If you are from a warmer climate and not accustomed to considering wind-chill, you may not immediately recognize ventilation equipment as a source that can cause wind-chill. However, think of it this way. A fan blowing air at a speed of 15 miles per hour is the same as a 15-mile-per-hour wind. If that manmade wind is blowing in an area where the temperature is 40°F, it produces a wind-chill equal to 22°F. You must manage exposure to cold temperatures in ways somewhat similar to high temperatures. Although they parallel those for preventing heat injuries—except that you are trying to warm the area instead of cool it—you should also follow these procedures:

- Change the temperature by heating the atmosphere within the confined space.
- Minimize the number of people exposed to the cold.
- Minimize the amount of time that personnel must wear impervious protective clothing, in this case to prevent sweating.
- Have rescue personnel pace their work to match conditions.
- Provide medical monitoring as conditions dictate.

SAFETY
Do not discourage efficiency, but rather monitor people so that they do not rush or take shortcuts that may injure themselves or others, or affect the rescue operation.

If you attempt to heat the atmosphere within a confined space, be certain it is safe to do so. You do not want the heater to become an ignition source. You also must ensure that the heat introduced into the space does not cause greater vaporization of any materials contained in the space. The vapors generated by the heat may be toxic, flammable, or both, and they may displace the oxygen within the space. If you know these conditions will not occur, you may want to consider if it is in the victim's best interest to warm the confined space or direct warmed air to the victim to protect him from hypothermia. Whatever you do, be realistic about what you can accomplish. Trying to boil water with the flame from a single candle is generally a fruitless activity.

Just as with exposure to high temperatures, you want to minimize the number of people exposed to the cold. Get people out of the wind, frequently relieve people who must remain in the cold to perform their assigned work, and rehab them in heated areas whenever possible. Try to keep both rescuers and victims dry and out of any precipitation that is occurring. When you are wet, your heat loss is increased tenfold. Cold temperatures and hot temperatures both will require additional personnel at the emergency scene. Protecting people against cold injuries benefits everyone at the incident.

When we discussed heat injuries, we pointed out that wearing impervious clothing such as firefighters' turnout gear or Tyvek™ suits are concerns because the PPE can cause sweating. Sweating is a cooling mechanism, and when people who are exposed to the cold sweat heavily under their protective clothing, they can rapidly lose body heat. Not something you want in cold conditions. If the work your rescuers are performing causes them to sweat and you cannot slow down the pace to prevent the sweating, you must consider how to either rotate the workers to get them out of the cold or how to get dry clothing for them.

As previously mentioned, pacing the work is always a good idea. It keeps people working within reasonable physical limits and can lead to greater efficiency. Pacing your work in cold weather is advantageous because of the layers of clothing you are wearing. Not only do you not produce as much body heat and not sweat as much, but layers of clothing affect movement and dexterity. There is another advantage to maintaining a modest work pace during cold weather. You also pace your rate of breathing. Each time you inhale in cold weather, you bring cold air into your lungs. At a normal pace, your body can warm the air and limit the cooling of the blood. When you begin breathing faster during cold weather, your body cannot warm the air in your lungs, which can lead to cooling of the blood and an increased loss of body heat. Match the rate at which you are working with the conditions you face. Accept that you are doing the best that you can under conditions that you may not be able to change.

Medical monitoring during cold temperature conditions, just as with hot conditions, is intended to recognize and prevent injury. In the case of cold temperatures, injuries include hypothermia as well as other cold injuries such as frostbite. Victims of hypothermia may not recognize that they have fallen victim. This failure to recognize the onset of hypothermia may cause them to deny that they are suffering from it. Just a 1-degree drop in core body temperature is enough to initiate hypothermia. Reducing your core temperature from 98.6°F to 96°F may be all it takes. Implement medical monitoring of vital signs, especially body temperature, to ensure recognition of hypothermia, provide proper treatment, and prevent more serious injury.

PERSONAL PROTECTIVE EQUIPMENT

Personal protective equipment (PPE) is equipment worn to provide protection to the body and/or its systems. This includes the head, arms, hands, legs, feet, torso, eyes, hearing, and respiratory system. Not all PPE is created equal, nor is it useful in all situations. Look at the variety of head protection available for different uses. A batting helmet is useful in baseball but would not provide adequate protection for a hockey player. Similarly, the head protection emergency workers wear at a confined space incident must meet the protective requirements needed in a confined space. Firefighters' helmets are designed to provide protection form the hazards present while fighting fires. They are designed to shed water and keep heat and debris off the firefighters' head and shoulders, among other things. As a result, the helmet is very large from side to side and front to back. That wide profile makes the helmet difficult to wear during entry into a confined space. A better choice of a helmet would be a low profile helmet such as a caving helmet or a hard hat with a strap, as shown in Figure 8-8.

> **NOTE**
> Firefighter helmets present certain problems at confined space emergencies.

Respiratory Protection

Respiratory protection for emergency use must be either a self-contained breathing apparatus or

FIGURE 8-8 Shown here (left to right) are a rock climbing helmet, a rescue helmet, a firefighter's helmet, and a hard hat with a strap. Notice the size of the helmets in relation to each other. If you expect rescuers to wear helmets, the helmets must be able to fit into the space and be worn without interfering with movement.

FIGURE 8-10 A supplied air respirator showing the air supply, air hose, escape bottle, and facepiece.

a supplied air respirator. What is critical here is that (1) the respiratory protection must have its own air supply and (2) the equipment does not interfere with entry into or exit from the space, as shown in Figure 8-9. You cannot use air purifying respirators or other equipment that relies on the air within the confined space as a source of breathing air. Both SCBA and SAR have limitations inherent to their design.

A SAR, shown in Figure 8-10, must have an escape bottle attached to it in the event of the

FIGURE 8-9 Depending on the size of the opening and the interior dimensions of a confined space, different people will be better suited for entry. A very small opening or low head clearance will limit who can work within the space.

failure of the airline or the external air supply. Typical escape bottles last for 5 or 10 minutes. SAR must also be connected to the external air supply by a hose. There are limitations to how long the air supply hose can be, and the supply pressure for the hose must be regulated to ensure that the user is getting the proper flow of air. The external air supply must also be monitored to ensure that adequate warning is given before the air supply runs out. When using a SAR, the air hose can be no longer than 300 feet. Longer hoses create too much resistance to the passage of air and can inhibit the proper positive pressure and liter flow per minute to the facepiece. You must also be concerned about maneuverability while using a SAR. You can have up to 300 feet of air hose to drag around within the confined space, and you must consider the air hose as you negotiate through obstacles in the space. Even with those limitations, SARs may still be your first choice for rescues. You will have an almost limitless air supply, there is an escape bottle that allows the user to disconnect from the air hose in order to leave the space, and the harness and escape bottle produce a low profile for entry and exit. This is what makes SARs valuable for confined space entry, especially when you are faced with entering tight openings. The key to successful SAR use is procedures and practice. Know how to inspect and maintain the SAR, how to don the equipment, and how to react to an emergency. Then practice until it is innate.

Just as SARs have limitations, so do SCBAs. There are significant limitations to the air supply because it is of a much shorter, finite duration (30-minute, 45-minute, and 60-minute rated). The size of the SCBA can make it difficult or impossible to enter the confined space when there is a narrow opening. That size factor might lead you to believe that you can temporarily remove the SCBA from your back, continue to wear the facepiece, breathe off the unit and enter the space. If you remove the SCBA to fit through the opening, you will need the acrobatic abilities of an ape. Chances are you are probably going to get into the space holding your SCBA in one hand, drop it, and pull the facepiece off. All you have done is increase the number of victims by one. You either can enter wearing and using the respiratory equipment or you can't.

> **SAFETY**
>
> You may conclude that all you have to do is temporarily remove the SCBA from your back while still wearing the facepiece and breathe off the unit. This conclusion is dangerously wrong.

If you make the decision to use respiratory protection, because of the hazards present, you should generally limit your selection to SCBA or SAR. The concern is that since you have a known hazard, the potential exists for an uncontrolled release of the contaminant or a failure of ventilation that will create an atmosphere that exceeds the ability of the air purifying cartridges to filter out the contaminant. Getting people into the space while wearing respiratory protection is a strategic factor.

Once the rescuers have entered the space, they may have to negotiate around other openings and restrictions within the space. These too are strategic factors, and you may not know about them until entry is made into the space. Your rescue entry team does not remove their SCBA once they are in the space. Just as with removing the SCBA during entry, the facepiece can get pulled off, and the rescuer may very well be taking his last breath.

> **NOTE**
>
> If you drop your SCBA and the facepiece is pulled off while you are in the space, you could very well be taking your last breath.

There are two basic types of respiratory protections—SAR and SCBA—and each has its limitations. If all you have is SCBA for use by your rescue team, do not try to make them fit an incident that requires SARs. Accept that the equipment limits what you can do and how you can do it. Either change the conditions or change your tactics.

> **NOTE**
>
> Either change the conditions or change your tactics.

Any rescue team that is serious about confined space rescue must have the proper types of equipment on hand or have it available for later use. If the equipment is available for use but it must be brought to the scene, how much time will that add? Can you take defensive actions in the meantime? If you are preplanning confined spaces and realize that the situations warrant having SARs with you as you arrive, consider purchasing the SARs, escape bottles, and any other SAR-related equipment for the rescue team to use. With or without preplanning, your response to confined space incidents will not be limited to only those incidents when SCBA will work. You are going to have incidents for which the solution is SAR, and you may have rescuers who insist that they can enter by removing and re-donning the SCBA inside the space. If they have to take off the SCBA to enter, put it back on, and then become a victim, how do they expect to be removed through the opening? If they were too big to fit in, they are too big to come out, no matter how hard the backup team pulls. Either you fit through the opening with the SCBA in place and in operation (this is where smaller people become more valuable) or do not enter the space!

Retrieval Equipment

In the event of an on-scene emergency, you must be able to rescue the rescue entry team that enters the confined space.

Harnesses

Rescue entrants must wear one of the most basic confined space pieces of equipment—a harness. It also must be the correct harness for the job. To help you select the correct retrieval equipment, The **National Fire Protection Association (NFPA)** has a standard for rope (NFPA 1983 Life Safety Rope) that also addresses harnesses. These harnesses are classified as Class I, Class II, or Class III harnesses depending on how they are intended to be used. The NFPA identifies the harnesses as follows:

- **Class I harnesses**
  - Designed to fasten around the waist and thighs, or around the waist and under the buttocks
  - Support a one-person load
  - Are not designed to hold you in the harness in the event you flip upside down (invert)

- **Class II harnesses**
  - Fasten around the waist and the thighs
  - Designed for rescue use when a two-person load may be encountered
  - Are not designed to hold you in the harness in the event you flip upside down (invert)

- **Class III harnesses**
  - Similar to Class II harnesses
  - Added advantage of straps that go over the shoulders
  - Provide protection from slipping out of the harness in the event that the rescuer is inverted by accident

Class III harnesses, shown in Figure 8-11, are the choice for confined space rescue. Not only do they address the potential for inversion, but they can be made with additional attachment points so that the rescuer can be lowered more easily. Additional attachment points can be located at the chest, upper back, shoulders, and the sides of the waist if needed. These additional attachment points allow a rescue entrant to be passed through restricted openings because they can keep their hands positioned (such as overhead) to provide the lowest profile. With theses additional points, rescuers remain in an upright and stable position, without having to use their hands.

FIGURE 8-11 This is a Class III harness with "D" rings at the shoulders and center of the back.

> **NOTE**
> Class III harnesses are the choice for confined space rescue not only because of the potential for inversion, but also because they can be made with additional attachment points that allow the rescuer to be lowered more easily.

At times harnesses are selected based on personal preferences, how they are constructed, or how they feel while being worn. This may be an acceptable practice as long as the harnesses are designed and manufactured to meet rescue requirements and the corresponding standard. When you require your equipment to meet a recognized standard, the equipment you purchase will perform as expected and is designed for the intended use. In addition to NFPA, various organizations create standards, including the **Occupational Health and Safety Administration (OSHA)**, the **California Division of Occupational Safety and Health (CALOSHA)**, **Underwriters Labs (UL)**, and the **American National Standards Institute (ANSI)**. Some of the organizations that create standards are consensus organizations (that is, their members develop standards and then vote to adopt the standards), whereas others are government agencies that both develop and enforce the standards. Very often government agencies such as OSHA adopt the recognized consensus standards by reference. Standard-making organizations do not certify that any of the equipment manufactured to their standard meets the particular standard or criterion. That is the job of an independent testing laboratory that uses the information in the standard to certify that a representative sample of the equipment has been tested according to the standard and meets the requirements. Different organizations may have similar standards for the same equipment. NFPA and ANSI try to maintain corresponding standards, but because the revision and adoption of standards can vary between organizations, it is important that you verify the applicability of the standard to your use of the equipment. Make sure that the way you will use the equipment matches the intended use in the standard. An example of this is SCBA for use in industrial applications as opposed to SCBA for use by firefighters. There are different requirements for how many liters of air per minute are delivered to the wearer, whether the straps will melt, and other features you may not be aware of. SCBA designed to meet NFPA's standard is based on use by firefighters and the demands of their work. The conditions this equipment is used under are different, and the amount of work a firefighter (or other rescuer) performs while wearing SCBA is significantly more physically demanding than work that a person wearing SCBA for general industry use would perform.

The standards you use to specify your equipment must be current. NFPA routinely updates its standards. OSHA first adopted many of its standards in the late 1970s and some have remained in place, unchanged, since that time. Even when government agencies adopt new standards, the version that is adopted may not be the most current version. When you use the appropriate standard, you ensure that the equipment you purchase will perform and can be maintained in a reliable fashion. You can expect the equipment to work as designed, and by specifying that recognized standard you can compare equipment prices and features. Using a standard eliminates guesswork and delivers you the right equipment for the job. Though you must meet the minimum enforceable standard for your equipment that does not prohibit you from exceeding minimum standards. It is all about what you want the equipment to do and, when people's lives depend on that performance, how you manage the risk.

Gloves

Wear gloves to protect your hands, and wear the proper gloves. If you are grasping ropes, wear gloves that give good grip and protect against rope burns. If you have to worry about sharp objects, leather gloves may be adequate, or the sharp object might be able to penetrate the leather glove or cut it. In that case, get gloves that are puncture- or cut-resistant. Wear gloves that provide not just the protection you need but that allow you to work in them. If you can't tie a knot or operate a buckle with the gloves on, chances are you will take them off. Firefighter gloves made to NFPA Standard 1973 are designed for firefighting—they keep out heat and water and are resistant to punctures and cuts. But these gloves are of limited use in a confined space rescue because they are generally bulky, difficult to use when tying knots, and do not provide for a good grip when handling ropes and cables. In confined spaces, team members should wear gloves that allow them to operate and use the equipment and still provide adequate protection from the hazards, as shown in Figure 8-12.

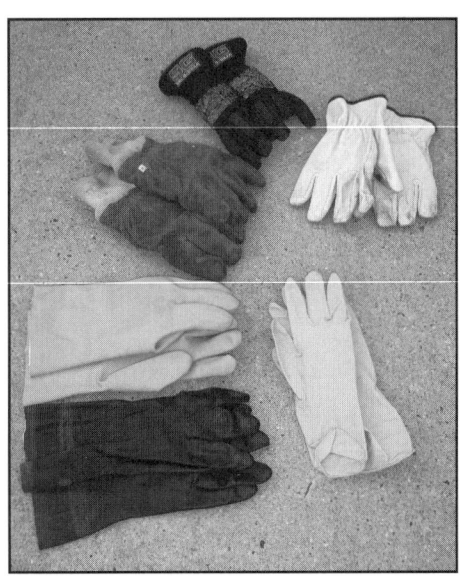

FIGURE 8-12 Shown here (top to bottom) are chemical-resistant gloves, firefighter's gloves, leather utility gloves, and rescue gloves. Each type of glove has a purpose and limitations.

Any discussion of gloves for confined space rescue must include latex or vinyl gloves for infection control. These gloves are designed to prevent the transmission of certain diseases while you are working on a patient. Depending on the circumstances of the incident, you should expect to come into contact with the victim's bodily fluids. Victims may also be contaminated from the space they were trapped in. A victim simply stranded in a sewer and not bleeding or otherwise losing bodily fluids may still be contaminated by waste products from the sewer, the equipment, or the area. Protect yourself. When necessary wear latex gloves under your leather gloves and properly discard all gloves that have been contaminated.

> **NOTE**
> Wear a pair of leather outer gloves to protect your hands.

> **SAFETY**
> Consider whether it is necessary to wear latex gloves while performing confined space rescues.

> **NOTE**
> If contamination is suspected, the outer gloves must be decontaminated or discarded.

Foot Protection

Foot protection is a fairly simple matter. Foot protection needs to protect your feet from toe injuries, puncture wounds, cuts, abrasions, liquids, and any other hazard that is present, while still providing traction and maneuverability. It may sound like a good pair of steel-toed work boots would fit the bill, but that may not be the case. Rubber firefighters' boots are waterproof, but they may not resist chemical contamination or provide adequate traction or maneuverability. So how do you select the right boot? This is when preplanning or a good size-up pays off. If you know chemical contaminants are present and that they could penetrate the boot, then match the foot protection to the hazard. That does not mean you have to buy special boots for each location, but it does means you have to have some method of protecting the boots you normally wear. Based on the hazardous material you expect to come into contact with, that protection may be as simple as a protective over boot or cover, or boots selected specifically for compatibility with the haz-mat (for example, polymer, PVC, neoprene). If contaminants are present, try to take steps to avoid contact with them. It is better to prevent contact with the contaminants than to have to go through decontamination. When that is not possible and your boots (or other PPE) become contaminated, either decontaminate the items or dispose of them. It may be more effective to purchase inexpensive, disposable, chemical-resistant boots and have them on hand for emergency use, than to replace firefighter boots. The type of boot you select and use will be driven by your local requirements.

Skin Protection

The skin is the largest organ of the body and is susceptible to damage by mechanical injury (for example, cuts, abrasions, punctures) and contact with chemicals. Additionally, the skin can absorb heat from or radiate heat to the surrounding environment. Identify the hazards to the body (arms and legs, too) and protect the skin. When you are looking at these hazards, make sure you look at the entire emergency scene. Cold weather affects not only the rescue team, but support personnel as well. You do not want to have your incident command team suffer frostbite or serious sunburn just because someone forgot to think about such injuries. Protection from heat or cold requires equipment that shields the body by either insulating it or reflecting heat away from it, or by warming or cooling it. An incident that requires specialized thermal protection (for example, steam suits, proximity clothing) may be highly specialized and well beyond your capabilities. Gloves, knee pads, elbow pads, and similar protective equipment can provide some protection against mechanical hazards. Mechanical protection should be worn when the space has unusually sharp edges, as shown in Figure 8-13. The goal is to prevent punctures and cuts—usually to the arms, hands, and legs, because these parts of the body move frequently and make contact with the hazard.

FIGURE 8-13 Sharp edges such as the weirs shown here require responders to protect themselves from cuts and puncture wounds.

The selection and use of chemical protective clothing at a confined space rescue incident should follow the same criterion used for hazardous materials incidents. The different levels of PPE are designed to provide appropriate levels of protection against skin contact so that you can avoid contaminants directly contacting your skin or absorbing into your skin. Contact by itself can cause serious injuries to the skin. Contact and subsequent absorption of the chemical can cause injury as the chemical reaches the blood, enters the bloodstream and affects the internal organs.

The **U.S. Environmental Protection Agency (EPA)** categorizes protective clothing by using four levels of protection based on the extent to which it protects the skin and the respiratory system. These four levels are A, B, C, and D.

Level A PPE The highest level of protection is **Level A PPE.** It provides both skin and respiratory protection, including protection from hazardous chemical vapors (via skin contact or inhalation). To protect from vapors, the Level A suits are gas-tight suits, and the wearer uses either an SCBA or a SAR and an escape bottle worn underneath the suit. Wearing Level A protective clothing for confined space rescue is extremely difficult. Imagine putting on the suit, having a harness put on outside of the suit, then making entry through the opening to the space. With an SCBA or SAR in place and the suit closed, how can you put a harness over the suit and adjust it so that the harness is wearable (not just comfortable, but usable)? The harness over the Level A suit may also prevent you from being able to access your respiratory protection in an emergency, because it is under the suit and the harness, and you may not have access to the emergency bypass. Add to all of that needing to maintain the integrity of the suit (you need to keep the suit gas tight) and not wanting to mechanically damage the suit. Wearing Level A protection is not impossible, but it is very difficult and greatly increases the overall risk level.

Level B PPE The next level of protection is **Level B PPE.** This is a splash suit with the SCBA or SAR worn outside of the suit. Level B protection allows you to don the protective clothing, put on a harness, and then put on the SCBA or SAR over that. Though this method is easier to use, it also provides less skin protection. You may be protected from contact with liquids and solids, but your skin is not protected from vapors.

Levels C and D PPE **Level C PPE** and **Level D PPE** are normally not used for rescue entry into a confined space that presents a respiratory hazard. Level C is similar to Level B because both are splash suits, but Level C has an **air purifying respirator (APR).** Air purifying respirators provide limited protection against inhalation of chemical vapors, and you must know the concentrations of the chemical vapor and not have the potential for an uncontrolled release that could increase the concentration. APRs do not provide protection against oxygen-deficient atmospheres because their source of air is the surrounding environment. Level D provides no skin or respiratory protection because it no more than ordinary street or work clothes.

Damage to Chemical Protective Clothing

Chemical protective clothing can be compromised by three different methods: permeation, penetration, and degradation. Once it is compromised, chemicals can enter the suit and make contact with the skin.

The movement of a chemical through the protective clothing on a molecular level, as shown in Figure 8-14, is known as **permeation.** When

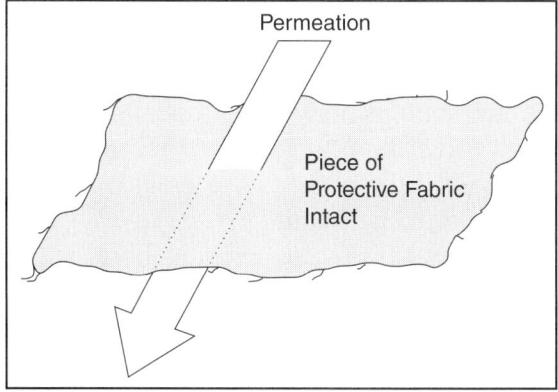

FIGURE 8-14 Illustration of permeation.

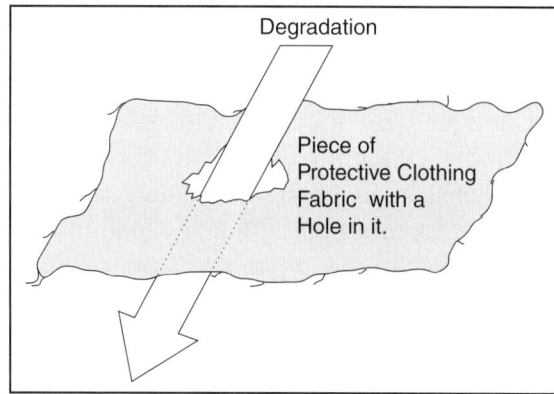

FIGURE 8-16 Illustration of degradation.

permeation occurs, the chemical moves directly through the fabric of the protective clothing and may not leave any visible damage. The resistance to permeation and the rate or speed at which it occurs will depend on the chemical and the fabric of the protective garment. No one fabric resists permeation by all chemicals. Your selection of protective clothing must be based on the chemical involved and on the ability of the protective clothing to resist permeation.

Penetration occurs when the chemical can move through openings in the protective clothing, as shown in Figure 8-15. Penetrations may occur through zippers, flaps, or other openings through which a person enters the suit. Penetration can also occur through seams and other openings that are created when the suit is manufactured. Based on the characteristics of a particular chemical, the hazards it presents, and the physical state (solid, liquid, or gas), the type and number of penetrations will affect the performance of the suit. To seal these potential spots for penetration, suits may be manufactured with taped seams or gas-tight zippers, such as in a Level A suit. You may be able to minimize the effects of penetration, on the scene, by duct taping zippers, cuffs, and other openings.

Degradation of the protective equipment is caused by tearing, dissolving, or wearing away of the fabric, as shown in Figure 8-16. Contact with a chemical may cause the equipment to degrade to the point that it begins to come apart due to the type of fabric it is made out of or because fasteners such as glue and thread begin to dissolve. Tearing can be caused by mechanical stresses on the fabric or seams, or the suit fabric may become worn through as it is brushed against a rough or sharp surface.

No matter how the protective equipment can be compromised, no single chemical protective fabric can handle all chemical hazards. You will either have to identify the chemical you will be coming into contact with and select protective equipment that resists the chemical, or assume a worst case scenario and go with the highest level of protection while you analyze the hazards. As we stated earlier, preplanning can provide big dividends in terms of safety and timely operations at a rescue.

If you are using chemical protective clothing in anticipation of being exposed or potentially exposed to a chemical hazard, you must also use a decontamination procedure to remove or reduce the contamination to acceptable levels before removing the clothing. Sometimes your best decontamination efforts will reduce the chemical hazard only to the point that people can safely remove their protective equipment. No additional decontamination is practical, and the equipment must be properly disposed of. When you are faced with an incident that involves a chemical hazard,

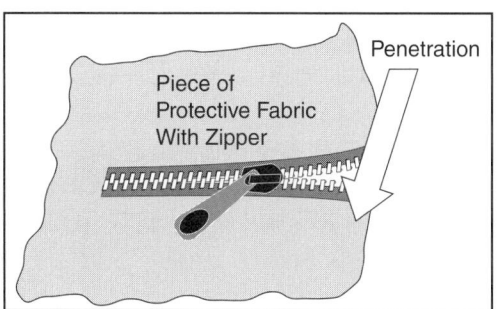

FIGURE 8-15 Illustration of penetration.

your level of training and experience will dictate what you can do. This book is not meant to provide detailed information about hazardous materials. There are very specific standards that identify the training requirements for hazards materials emergencies. In advance of an emergency, define the level of haz-mat training and capabilities your confined space rescue team has. Stay within those limits. If the operation is beyond your capability, call for help.

> **NOTE**
> When you are dealing with a chemical hazard, your own level of training and experience will dictate your capabilities.

NOISE

If you were asked to identify significant hazards at an emergency, would noise be one of the hazards you would name? Many times we consider noise as part of an emergency. There are sirens, running engines, alarms, radios, and lots of other sources that create noise during an emergency. At times the noise causes confusion or difficult communication. Confined spaces can increase the effects of noise and may require hearing protection. Because of the shape, construction, or configuration of a space, the noise within the area may be amplified by the space itself. The old expression that "an empty barrel makes the most noise" illustrates this point. Confined spaces can behave like empty containers and allow noise to echo within the space. Often nothing is in the space to absorb the noise. Trying to communicate at a normal volume while working in a large bulk storage tank can be difficult because the sound echoes around the tank and it becomes so loud that it seems people are shouting. You may also find that the echo disrupts the sound of the person speaking and, even though you are nearby, you cannot understand what is being said. Whatever the cause of the communication failure, noise is affecting your operation. That inability to communicate presents an acute danger because you may not be able to speak with entry team members or warn them of danger. How can you tell the entry team that you will be changing the action plan if they are in the space and cannot hear or understand you? How can you order them to evacuate because of some new danger if the noise is too great for the evacuation order to be heard?

Amplified noise also creates both an acute danger and a chronic risk to the people exposed to it. Sound is measured on a scale in relation to the intensity of sound. The unit of measure is a **decibel**, which measures the pressure of sound. The more decibels, the greater the sound pressure and the greater the potential for injury from the sound. Sound can create so much pressure that it is painful. Sound pressure is managed through hearing protection, which can be handled in several ways. Eliminating the sound is an obvious method to control it. Tuning off equipment, stopping unneeded noise-producing work, and creating quiet tactical areas as needed can reduce the noise. But it may not be possible to eliminate the noise completely. When you cannot eliminate the sound, you need to try and reduce the level of the sound as much as possible. Sound reduction can take place in several ways. Sound-absorbing materials can dampen the sound at its source, but that may not be practical at an emergency. At an emergency, you should rely on hearing protection equipment worn to reduce the number of decibels that reach the ear. Hearing protection equipment can consist of earplugs or special earmuffs. Recognize noise as a hazard that needs to be controlled at an emergency, identify the source of the problem, and solve it. No matter what the source, noise that interferes with communication must be dealt with. When verbal communication is affected, you may have to use visual signals or specific sound signals such as sirens or air horns for emergency evacuation alerts. Noise that affects how people hear requires the use of hearing protection.

> **NOTE**
> Hearing protection can consist of earplugs or special earmuffs.

LESSONS LEARNED REVISITED

The victim, firefighter, and paramedic all were victims of heat stroke. The weather at the time of the incident was overcast with a temperature of 90°F, high humidity, and little wind. The tank was 12 feet deep and had very little air movement for cooling, and the bottom of the tank was black because of the coal. This tank is a confined space.

Discussion Questions

1. At what point would you initiate air monitoring?
2. If the workers were still in the filtering tank when you arrived and police, fire, and EMS personnel entered the tank as they arrived, would you consider that safety was being properly addressed?
3. Because you have what appears to be at least three people suffering from heat stroke, would you expect that other people at the incident may also be suffering from some sort of heat stress?
4. How would you prevent that potential injury?

SUMMARY

- The incident commander has the primary responsibility for safety.
- Rescuers cannot assist victims if they become victims themselves.
- Addressing safety issues for emergency response begins before the response.
- Selecting, using, and training with equipment is part of safety.
- Protect people against heat-related injuries, cold, and hypothermia.
- Use PPE when needed and make sure you are using it correctly.
- Safety must be part of the culture of the emergency organization.
- Identify hazards and eliminate them, or protect people from the hazard.
- Safety is a strategic factor.

REVIEW QUESTIONS

1. Describe the conditions (temperature, humidity, etc.) under which heat stress injuries can occur.
2. Describe the conditions under which hypothermia can occur.
3. Identify how temperature-related injuries can be prevented and the effects those prevention measures can have on your rescue operations.
4. NFPA Standard 1983 covers life safety rope and related equipment. Describe the value of using equipment that has been designed and built to nationally recognized standards (for example, ANSI, NFPA, OSHA).
5. Describe the purpose and limitations of at least three different pieces of protective clothing.
6. The movement of a chemical through protective clothing on a molecular level is _____.
   a. Penetration
   b. Degradation
   c. Permeation
   d. None of the above
7. Hearing protection is not needed at a confined space emergency. True or False?
8. It is an acceptable practice to enter a confined space by removing your SCBA and

then placing it back on after you have entered. True or False?

9. Using the wind chill chart in Figure 8-7, determine what the resulting temperature effect would be if the temperature was 40°F and the wind speed was 30 mph.

   a. 37°F
   b. 22°F
   c. 13°F
   d. 0°F

10. When the potential for heat stress injuries is present, the entry team should be required to hydrate only before entry. True or False?

KEY TERMS

medical monitoring 113
hypothermia 114
National Fire Protection Association (NFPA) 119
Class I harnesses 119
Clss II harnesses 119
Clss III harnesses 119
CALOSHA 119
American National Standards Institute (ANSI) 119
U.S. Environmental Protection Agency (EPA) 122

Level A PPE 122
Level B PPE 122
Level C PPE 122
Level D PPE 122
air purifying respirator 122
permeation 122
penetration 123
degradation 123
decibel 124

ACTIVITIES

1. Take one of your prior tabletop exercises or case studies, reevaluate it, identify only safety issues, and determine how you can address those safety issues.

2. Consider using prior exercises or case studies and change incident conditions (for example, weather, presence of hazardous materials, access), and identify how these changes can affect safety.

ADDITIONAL RESOURCES

Everyone Goes Home®, http://www.everyonegoeshome.com
16 Firefighter Life Safety Initiatives, http://www.everyonegoeshome.com/initiatives.html

OSHA Quickcards, http://www.osha.gov/pls/publications/publication.AthruZ?pType=Types&pID=6
OSHA Publications, http://www.osha.gov/pls/publications/publication.html

9 Rescue

LESSONS LEARNED: CONTRACTOR KILLED AND TWO INJURED IN A LIFT STATION ACCIDENT

A pipe fitter was killed at a storm sewer pumping station when an inflatable plug is being used to hold back water in a 26-inch storm sewer pipe burst. The pipefitter was part of a 5-person work crew and had entered the 10 feet wide by 10 inches long by 30 feet deep manhole with another worker to replace a stationary pump. The workers were at the bottom of the lift station when the plug exploded, collapsed part of the wall, and filled the station with 15 feet of water. Members of the work crew outside the lift station called immediately for help. Upon arrival, the fire department was directed in by the workers' foreman, who was injured, where they found two workers administering first aid to an injured colleague.

The site, a city-owned lift station, was familiar to rescue crews because they had used this location for training in the past. The lift station is used to lift storm water runoff collected from the nearby park and pump it in the river.

Critical Thinking Questions

1. If you responded to this incident, who would be in charge?
2. How many victims are there at this incident? Where are they?
3. What hazards do you anticipate at this emergency?

LEARNING OBJECTIVES

NFPA 1006, *2008 EDITION*
By the end of this chapter, you should be able to:

- Manage site operations including initial scene operations (5.2)
- List the nine-step process for confined space rescue
- Identify the different types of confined space rescue equipment available to the confined space rescue team and the limitations of the equipment (5.2.2)
- Categorize operations as defensive or offensive (5.2.2)
- Recognize the basic considerations you must have for victim assessment, victim stabilization, and victim removal (5.3.3)

INTRODUCTION

Your ESO has been called to a confined incident where three victims are trapped. You arrive on the scene and see the confined space opening with police, fire, and EMS workers looking into it. As you get out of the rig, other rescuers turn to your team and an EMS officer says, "Glad you are here. What should we do?"

RESCUE CONSIDERATIONS

This is it. Everything you have trained for will now come into play. Where do you start? What should you do first? The questions begin to flood through your head. The lives of three people depend on what you do next. You begin running a mental checklist of all the considerations you have been trained to recognize, and you begin sorting them out. Should you begin by ventilating the space, or should you use a meter to evaluate the atmosphere in the space? You chose to become part of the emergency services, and you did it because you value the chance to help people. Now you will have that opportunity. There has to be a jumping off point, but where is it? Let's start by breaking it down into steps that take you from the beginning to the end. To do that, you need to follow a simple nine-step process. This process is shown in Skills/Procedures Box 9-1.

We'll now discuss each of these steps in more detail.

Establish Command and Take Control of the Incident

Someone has to be in charge of the emergency. This is the requisite step for all other activities. If no one coordinates, directs, and evaluates the rescue operation, it will be difficult to do anything to assist the victim(s) the rescuers will be endangered. Without someone in command, as shown in Figure 9-1, your operation will be disjointed and stagger along. Any other part of the nine-step process cannot take place effectively. Have you ever been to an emergency at which lots of people were giving orders and no one was following them? Generally nothing gets accomplished under these conditions, except maybe for an increase in volume from the people giving orders. Your task at this point is to make a difference for the victim, but you can't do that

SKILLS/PROCEDURES 9-1
Nine-Step Rescue Process

1 Establish command and take control of the incident.

2 Identify the type of rescue problem.

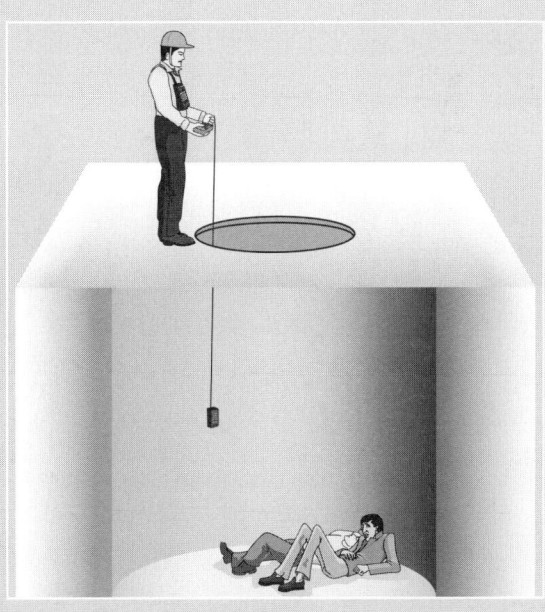

3 Perform hazard and risk assessment.

4 Identify rescue objectives.

SKILLS/PROCEDURES 9-1

Nine-Step Rescue Process

5 Identify resource needs.

6 Develop an action plan.

7 Implement the action plan.

SKILLS/PROCEDURES 9-1

Nine-Step Rescue Process

8 Evaluate the effectiveness of the action plan.

9 Terminate the incident.

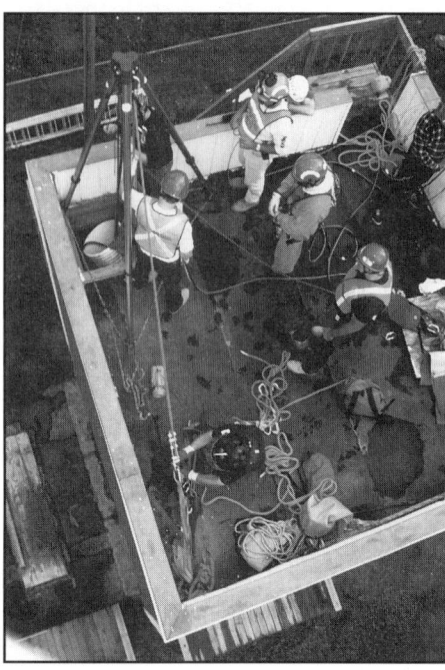

FIGURE 9-1 Effective control of the rescue begins with establishing command. Failing to establish command will affect all additional operations.

FIGURE 9-2 Identify the rescue problem. This is a discharge pipe for a storm water collection system. Multiple pipes feed to this pipe, and the victim could be anywhere within the collection system.

if the emergency operation falls apart. When the IC takes charge, tasks are coordinated, the action plan is sequential, and you are doing the best you can for the victim. You are the professional; people will look to you to take control, lead the operation, and do what you can for the victim.

Identify the Type of Rescue Problem

Is this incident a confined space accident, or is it some other type of rescue you have been called to? Just because you were told on dispatch that this was a confined incident doesn't automatically make it a confined space rescue. Sort out the problem, identify the type of rescue scenario, and go from there. Size-up for *some* confined space incidents begins with preplanning, but *all* confined space incidents go from generic to specific on dispatch. The dispatchers are given certain information by the caller and then relay that information to you. Based on the initial information you were given, you should begin a mental size-up of the emergency. On arrival, get more information and identify the type of rescue you are faced with. If you were dispatched to an incident with a report of a hand seen extending out of a storm drain and people screaming from inside a culvert like the one shown in Figure 9-2, how much more information would you need before you made entry? Match your capabilities against the tasks it will take to complete the rescue. If you have been dispatched to what was reported as a confined space rescue and find on arrival that it also requires high-angle rescue skills or a haz-mat team, you may find that you are limited in what you can do and need additional assistance. Last of all, confirm that a rescue is needed. When you arrive on the scene, the victim may already be out of the space.

Perform a Hazard and Risk Assessment

Hazard and risk assessment is an in-depth continuation of your size-up. You are not just looking for strategic factors, you are analyzing them and weighing the effect of each on your operations. Get the site-specific information such as confined space permits (as shown in Figure 9-3) and question witnesses to the accident or the people who discovered the problem. Use your monitoring equipment to evaluate the atmosphere in and near the space. The information you develop during the hazard and risk assessment is meant to identify hazardous materials within the space, oxygen-deficient or oxygen rich atmospheres, flammable atmospheres, and any other significant

FIGURE 9-3 Hazard and risk assessment begins upon dispatch to the incident and is an ongoing process during the incident. Here the rescue team is reviewing the confined space entry permit.

FIGURE 9-4 Identify your rescue objects. Is it necessary to enter the space to perform the rescue?

hazard that can be identified. What you do from this point forward will be based on the limits you have identified. Those problems that can be easily corrected or controlled and reduce risk are the simple problems. The more difficult problems will take more effort to manage.

Identify Rescue Objectives

For discussion purposes, let's say that your initial hazard and risk assessment shows that there is an oxygen-deficient atmosphere, the rescuers will have to wear SCBA, the space must be ventilated to protect the victim, and continuous monitoring of the space must occur. You now have several pieces of information that will require you to make some decisions. Based on the information you have and the hazard and risk assessment, you can see how the critical factors will begin to lead you in a particular direction. If you were to identify something especially critical, such as the presence of a hazardous atmosphere, your decision making should be so straightforward that you quickly identify the need to protect the victim, choose basic defensive actions and implement those actions while your team prepares to rescue the victim as shown in Figure 9-4. And you may need to do this while at the same time you protect other people in the immediate area. Focus on the incident priorities—life safety, incident stabilization, and property conservation—and the order of importance. How you implement them and the outcome you hope to achieve will lead you to your objectives.

The obvious goals are to successfully rescue the victim and protect the rescuers. But remember, goals are only broad statements of what you want to accomplish. Objectives are the steps you must take to achieve the goals. Objectives are the specific, measurable, and coordinated steps you will take to accomplish your goals. That being said, remember that your resources are limited, you can work only on accomplishing a finite number of objectives at a time, and the objectives should not work against one another. During your size-up and analysis, you should understand that achieving some objectives will have much more impact than achieving others. Sort out your objectives against two criteria: the difficulty you face in achieving them and the impact they will have on the emergency. Objectives that are easy to complete and high impact should receive immediate attention. These simple, high impact objectives might let you accomplish both life safety and incident stabilization at the same time and in a short time frame so that you protect the victim while buying time to complete rescue entry preparations. Difficult, low impact objectives should be avoided whenever possible. They may simply waste resources that could be better used elsewhere. Easy, low impact and difficult, high impact objectives may be the most difficult to accomplish. They can be resource intensive, may create additional risk to rescuers, and take time to accomplish. However, depending on the circumstances of the incident, they may be all you can do.

You cannot develop an incident action plan until you have set your objectives for the emergency. Even after you have set your objectives, realize that just because you thought a particular objective would be easy to complete and would be high impact does not mean that you can complete it. Evaluate what you are accomplishing against what you wanted to accomplish. Your understanding of the situation may not always be correct, and that misunderstanding can change your objectives.

Identify Resource Needs

Once you have identified what your objectives are, determine what resources you need to get the job done. Do you need additional people and specialized equipment, or is what you have on site adequate, as shown in Figure 9-5? Mange your resources, whether they are people or equipment. Chapter 5 talked about resource management and identified three categories for you resources: available, in use, and not available. There is more to resource management than knowing status. You must understand the limits of your resources. If you try to exceed those limits, you are building failure into your operation. If you are lucky, all you will waste is time. The more severe consequences of exceeding the limits of your resources can include damage to equipment or loss of that equipment, but more importantly, injury or death to the victim and rescuers. The more severe the potential consequences of that failure, the greater the risk you are taking. Resource management requires you to match your resources to the task, and it also requires you to manage the risk to which those resources are exposed. Allowing resources to be used beyond their design limits, failing to provide proper maintenance, or failing to train personnel on the proper use of a particular piece of equipment all increase risk. Know your resources, know their limits. and have a plan to get the additional resources you anticipate needing. People should be the number one resource you manage. Do not use them beyond their limits.

> **NOTE**
> The type of equipment used for confined space rescue will vary from simple to sophisticated.

Develop an Action Plan

This is the point at which you begin the transition from preparation to action. No one person or piece of equipment can accomplish a confined space rescue. It takes a coordinated plan to complete the rescue because there is interdependency between the different parts of your plan. Before you can enter the space, you may need to ventilate. Before you can ventilate, someone has to bring the fan to where it is needed. Simply saying it out loud—"Get a fan!"—will not accomplish anything productive. Someone has to be tasked with getting the fan, someone else needs to bring an extension cord if the fan is electrically powered, and someone else must find an electrical power source to plug in the extension cord. The IC is the person who develops the action plan, and they put it into action by directing others, no matter how complex the emergency. Sometimes it will be simple to put a plan into use. Sometimes the plan may be more complex and require simultaneous tasks to achieve the objectives, as shown in Figure 9-6. Without an action plan, people may start free-

FIGURE 9-5 What resources do you need, and what resources do you have available?

FIGURE 9-6 In developing an action plan, identify the sequence of actions needed to complete your operation. Equipment such as retrieval equipment may need to be in place prior to entry.

lancing and create their own competing plans. When you fail to plan, you plan to fail.

Implement the Action Plan

By this point in the emergency you know what you want to do, you know what you need for success, and you have created a plan. This is the point at which you have people execute the plan, as shown in Figure 9-7. Everyone should know their roles as well as how and when they are expected to perform them.

FIGURE 9-7 Implement the action. Once you are set up and ready to go, begin your operation.

FIGURE 9-8 Evaluate the effectiveness of your operation. How are your personnel, equipment, and other resources performing?

Evaluate the Effectiveness of the Action Plan

Review what is happening at the scene. Is it what you expected to happen? Or is it different? You must assess your progress, because action plans rarely run exactly as you intended. At times the plan will be effective and you will not need to make many adjustments. At other times nothing will seem to work, you will make constant adjustments, but you will struggle to complete the objectives you created. Evaluating the plan and its progress must be an ongoing process, as shown in Figure 9-8. Continue your evaluation of the effectiveness of your action plan until the incident is terminated.

Terminate the Incident

Completing your rescue or recovery operation is not the end of your operation. Depending on the circumstances of the emergency, there will be a list of things that must be completed to bring closure to the incident. At a minimum, you must inspect your equipment, perform any required maintenance, take damaged equipment out of service, and record required inspection, maintenance, and service information. That takes care of the equipment, but what about your people? You should debrief personnel to establish accurate information about the incident for incident analysis and critique, to improve the effectiveness of operations, to identify needed improvements for future incidents (as shown in Figure 9-9), and to have a factual basis for any

FIGURE 9-9 Debriefing rescue workers after an incident can provide valuable information for future use. These students are being debriefed after completing a training exercise.

investigation of the incident that may follow. If a serious injury or death occurred, government agencies will be investigating, as will as attorneys for both sides in any lawsuit that might follow. Recording the information at the time of the incident is much better than trying to remember it six months or a year later.

The debriefing is not just of value from a lesson learned prospective. It also provides an opportunity to present information to your emergency responders about extraordinary circumstances or conditions that may affect them. If hazardous materials were present at the emergency, you should advise personnel of the signs and symptoms of exposure to the haz-mat. You should also include symptoms that may be delayed, where to report such symptoms if they occur, and where to seek treatment. See if you can get an update on the condition of victims and let the rescue team know before they leave the scene. The news may be good or it may be bad, but the rescuers need to have accurate information, and they need to hear it from a single informed source.

Any debriefing must consider the need to address critical incident stress. Your personnel are people with emotions and feelings about what just happened. People will react to the incident, and they need to know that the incident is an abnormal situation. Their reactions to this abnormal incident are predictable, normal reactions. It is not they that are abnormal, but the incident.

Finding out that the victim has died or that the injuries are more serious than first thought can evoke strong emotions. The death or serious injury of a fellow emergency worker or a prolonged operation where you have little chance of rescuing the victim are also examples of situations that can lead to **critical incident stress**. If you are providing maintenance to your equipment resources, don't forget to provide maintenance for your human resources. As part of your team training, consider including recognition and awareness of critical incident stress. Identify critical incident stress teams that are available in your area, and know how to activate them and how doing so will benefit your people.

EQUIPMENT

The type of equipment you will use for confined space rescue can range from simple to sophisticated. The type of equipment you will need should be determined though preplanning. Simple equipment has the advantage of being easy to set up and use, it can require minimal training, and it can speed up a rescue. However, simpler is not always better. Whatever equipment you purchase must be right for the job. Think of the tools you use in everyday operations. To a firefighter, an axe is a symbol of the fire service in addition to being a valuable tool. The axe is valuable for opening doors, breaking windows, and opening walls, but it is seldom used anymore as the primary tool to cut a ventilation hole in a roof. Instead we see power saws being used to cut ventilation holes. Each tool (the axe and the power saw) has its place, and though you could use an axe to cut a vent hole, would you want to? The axe takes less training, always starts, doesn't need to be fueled, and doesn't require as much maintenance. But given the choice, which would you use for a roof? The same rationale holds true during a confined space rescue. Certain tools and equipment work better than others. But like the power saw, what maintenance do you need to provide? Not only must you provide maintenance to the mechanical parts, you must provide maintenance to the operator. Maintaining proficiency with the equipment is as important as having the equipment. People who do not know how to use the equipment or how to use it safely are dangerous.

Select the tools you will need for the job, train people to use them correctly, and maintain the capabilities of both the equipment and the people. Bringing those three items together will enable you to use the equipment properly.

So far in this book there has been discussion of equipment such as personal protective equipment, harnesses, blowers, and monitoring equipment. There is no need to repeat those discussions here, except to remind you that you must have that equipment available and know how to use it. What this portion of the book will address is the other equipment you may already have or will need to acquire for use at confined space incidents.

Tripods

A **tripod**, shown in Figure 9-10, is one piece of equipment that is closely associated with confined spaces. When you choose a tripod, you must be aware that there are differences based on the height of the unit, footprint, and lifting capacity. How you plan on using the tripod will affect the number of anchor points on the head of the tripod, attachment points for lifting devices, the adjustability of the height, and loads you can support or lift using the tripod.

> **NOTE**
> Tripods are meant to lift and lower people. Lowering or raising equipment or other loads into and out of a confined space with a tripod should be forbidden.

Tripods are one device typically meant to support people as they are raised or lowered into and out of a space. Because the tripod will support people, it should not be used to support any other loads. When you support a person on the tripod, its load capacity must not have been compromised by the potential overloading of equipment or prior stresses that could have caused unseen damage and may lead to unexpected failure. Properly loading the tripod is just as important as not overloading it. The tripod has an axis that runs through its center to initially pickup the load and it then transmits the load to the ground through the axis of each leg. To evenly load each leg, you must ensure that the load is applied along the center of all those axes. If you load the tripod in any other direction, the load will want to go in that direction. So if your load is at an angle to the axis or perpendicular to the axis, you can pull the tripod over. The greater the angle to the axis, the more unstable the tripod will become. To avoid tipping the tripod, you must use either equipment specifically designed for use with that tripod or an anchor point and a change-of-direction pulley to keep the direction of the load aligned with the tripod and then redirect the load using the change-of-direction pulley. This may sound like a simple process, but to achieve your desired outcomes, you must know the capacity and limitations of your equipment and how to properly use it. The equipment manufacturer is a good source of this information.

Even when you are able to use a ladder to enter a confined space, a tripod is valuable because you can set up the tripod, connect the rescue entrants to a retrieval device on the tripod, and create an entry team rescue device that can be used from outside the space. If you need to rescue the rescuers from the space, no one has to

FIGURE 9-10 Among the various pieces of equipment that can be used for confined space rescue is a tripod. Know the capabilities and limits of your equipment.

make entry or attempt to lift the injured rescuer out of the space. You simply let the equipment do the work for you. Another advantage to setting up and using a tripod at the same time you are using a ladder is that you can use the retrieval line for fall protection, or as a fall arrester, and limit the free-fall potential of a person climbing up or down the access ladder. When using a tripod as your sole means of entry and exit, each person entering the space should wear a retrieval line and a safety line while they are being raised or lowered. With two rescuers in the confined space, there should be two retrieval lines in use and a spare third line.

This requirement for the three lines (one retrieval line for each rescuer after they are in the space and a safety line for use during lifting) also requires three attachment points at the head of the tripod. Each line should be as independent of the others as possible. When it is time to remove the victim from the space one rescuer would have to exit the space using one retrieval line and the safety line. Both lines would then be sent back into the space to be used to in lifting the victim. Once the victim was removed, the safety line would be lowered back in for use by the second rescuer as they exit the space. Though this may seem redundant, the use of a safety line ensures that if anything happens to either the retrieval line or the hoisting equipment, you have a backup means to safely recover your personnel and the victim. This process is illustrated in Skills/Procedures Box 9-2.

> **NOTE**
>
> You can effectively use three lines with two rescuers and a victim. It is only a matter of sequencing the ropes and the victim removal. With two rescuers, the first rescuer enters with a retrieval line (line 1) and a safety line. The safety line is sent back up and the second rescuer enters with a retrieval line (line 2) and safety line. Once the victim is ready to be removed, the second rescuer is taken out of the space first using line 2 and the safety line. Line 2 and safety line are then sent back and connected to the victim, and the victim is removed. The safety line goes back in the space for rescuer 1, who is removed from the space.

> **SAFETY**
>
> Using a safety line seems redundant, but it is a simple way to ensure that if anything happens to one line or to the retrieval equipment on it, you can safely recover your personnel and the victim by using the second line.

FIGURE 9-11 This tripod has three safety/retrieval systems attached to it. If it is used to support more than one person at a time, will the tripod be overloaded?

Limitations to the effective use of a tripod should be considered when you are purchasing it. How long a cable do you need on the retrieval system? What is the maximum workload you anticipate? You should consider these items before you purchase a tripod. Also take into account other considerations such as the weight of the unit, the ability to adjust the leg height, stabilizing devices to keep the legs from spreading, and any other special need you may be aware of. Based on what you know about the rescue situations that you face, you should specify and purchase equipment that will fit your needs. Without a tripod or in a situation where you cannot use a tripod to raise and lower people for rescue purposes, you may be able to improvise the use of other equipment for use.

SKILLS/PROCEDURES 9-2

Managing Three Retrieval/Safety Lines for Two Rescuers and a Victim

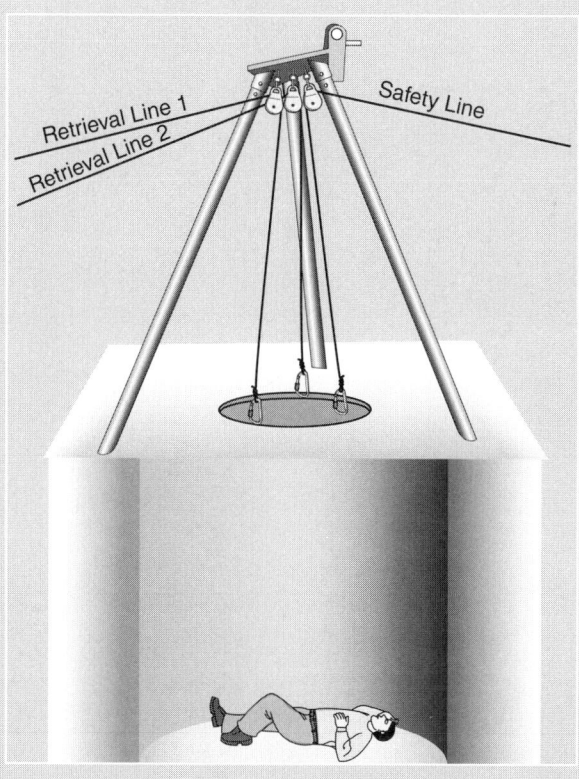

1 You have one tripod that accepts three attachments for retrieval/safety lines.

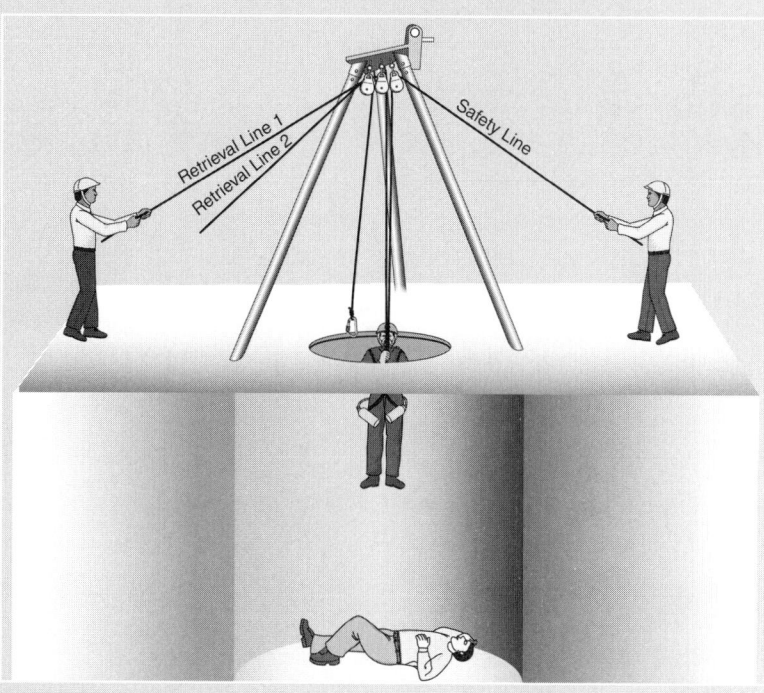

2 One rescuer enters wearing retrieval line 1 and one safety line.

SKILLS/PROCEDURES 9-2

Managing Three Retrieval/Safety Lines for Two Rescuers and a Victim

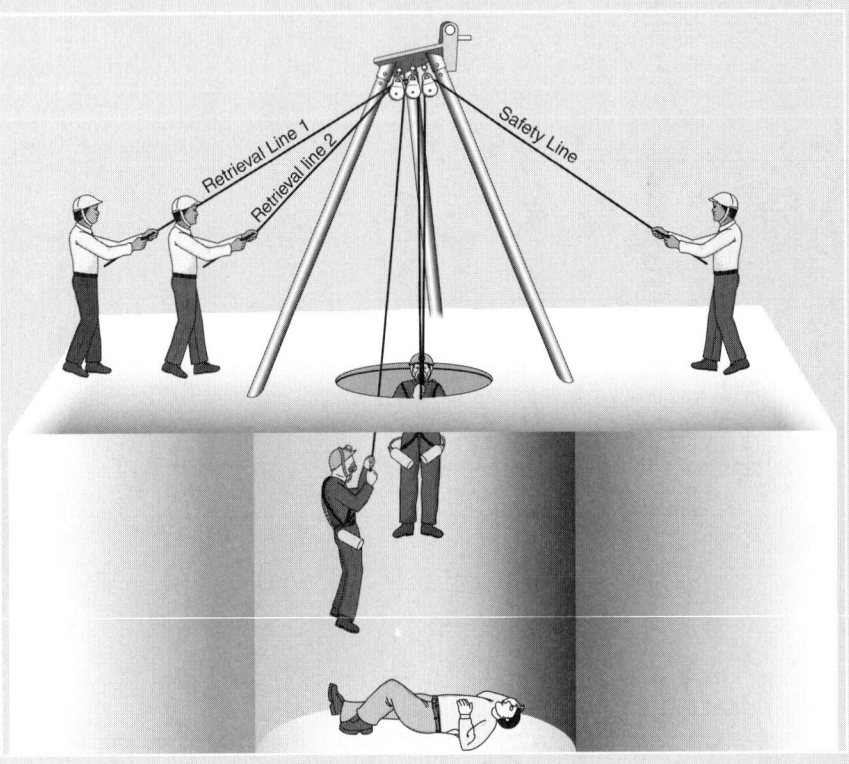

3 The first rescuer detaches the safety line, which is brought out of the space and attached to the second rescuer along with retrieval line 2. Rescuer 2 then enters the space.

SKILLS/PROCEDURES 9-2
Managing Three Retrieval/Safety Lines for Two Rescuers and a Victim

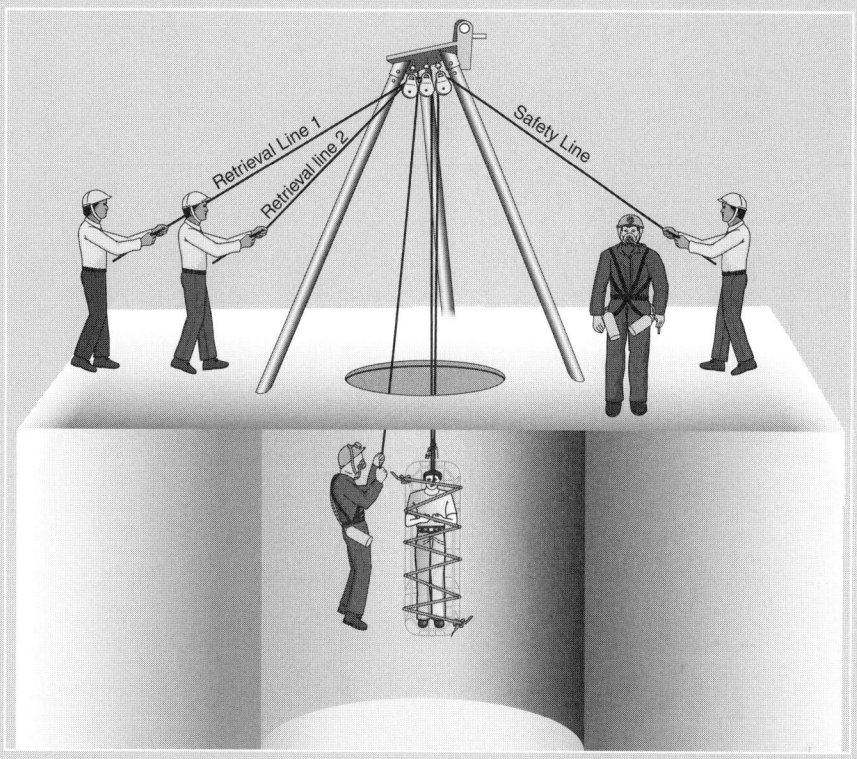

4 When the victim has been packaged by the rescuers, rescuer 2 is brought out of the space, and retrieval line 2 and the safety line are detached from rescuer 2 and sent back into the space to be attached the victim. Rescuer 1 attaches retrieval line 2 and the safety line to the victim, and the victim is removed from the space. Retrieval line 1 remains attached to rescuer 1 at all times.

SKILLS/PROCEDURES 9-2

Managing Three Retrieval/Safety Lines for Two Rescuers and a Victim

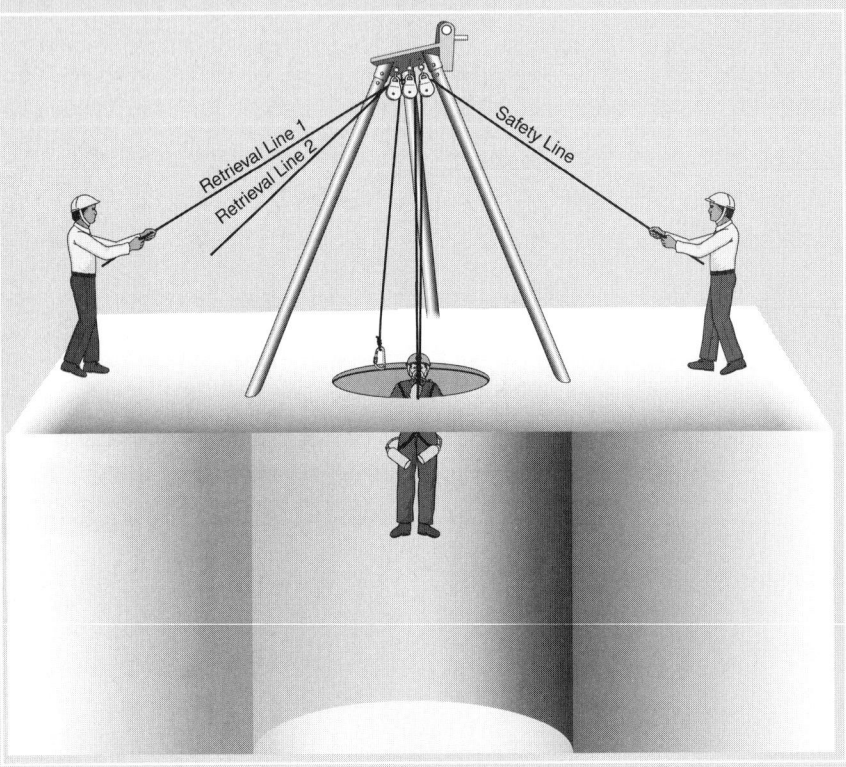

5 The safety line is sent back into the space, rescuer 1 attaches it to his harness, and he is removed from the space.

Improvising Lifting Devices

Using ground ladders to build A-frames or using aerial devices (for example, ladders, tower ladders, Squrts™) as a gin pole for lifting may be possible. However, you must absolutely know the design and working limits of the equipment you will be improvising. Ground ladders have very specific load limits, and they may or may not be capable of supporting the load you will place on them. How you fasten the ladders together will affect the capacity of the system you build. Cobbling two ladders together with an adequate rope of nylon strap provides an inherent weakness to the system. Don't try to support a 200-pound load with straps that will break at 50 pounds.

Aerial devices are designed to carry different loads. In the 2003 Edition of NFPA 1901, aerial ladders must have a minimum rated capacity of 250 pounds on the outermost rung with the ladder fully extended horizontally. There are minimum design load requirements for the rungs of an aerial ladder that are based on the weight of the load and the area over which the load is spread on the rung (500 pounds distributed over a 3-inch. wide area at the center of the rung). Other aerial devices have different design load requirements. All of these design standards anticipate an aerial device being used as it was intended, to fight a fire. Improvising the use of an aerial device should not be done without knowing its limits. Newer aerial ladders may be able to support larger loads at lower elevations, but what was your ladder designed to support? And remember one basic limitation of an aerial ladder: The load is meant to be applied equally between both beams. If you load one side more than the other, you create a torsional load. The ladder can twist and fail under a torsional load.

An aerial device is also a mechanically powered piece of equipment. Mechanically powered equipment can apply large forces and injure people. If a person were to become trapped or wedged during retrieval, the power of the aerial device might continue to be applied on the trapped victim's body That is why the use of mechanically powered hoisting devices should be limited to those devices that have a clutch that will limit the amount of force that can be applied. If you want to use improvised equipment, know the limits of the system you want to create, preplan, train using the equipment and the methods that you want to employ, and use the improvised equipment only after careful consideration. Inventing new and untested devices in the field is dangerous and should not be permitted. Talk to the manufacturers about the limitations and use of aerial equipment, ground ladders, and other rescue equipment. Get the design specifications for the equipment.

> **SAFETY**
>
> Mechanically powered equipment poses the additional danger of injuring people because of the forces it can apply in the event that a person becomes trapped or wedged while being lifted.

Rope and Equipment

Rope can be a valuable rescue tool at confined space rescues, as shown in Figure 9-12. This book is about confined spaces and their unique problems, so the discussion on ropes and associated equipment will be brief. Rescuers must have a basic understanding of rope terms and the types of rope that are available. Rope can be made in several different ways, including laid ropes, braided ropes, and kernmantle ropes. **Laid ropes** have twisted strands, and all of the strands contribute equally to the rope's strength. **Braided ropes** are made of strands that are braided together, and these ropes may have an

FIGURE 9-12 Know what your equipment is designed for and how it is meant to be used.

outer jacket. **Kernmantle ropes** use a core called a kern that is protected by an outer jacket called a mantle. The kern or core carries up to 75 percent of the load placed on the rope, and it is protected by the mantle.

Ropes can be static or dynamic. Static rope has little stretch and is typically the choice for rescue. Dynamic rope is designed to stretch and is often used when shock-absorption is required, such when there is a danger from a fall. Ropes can be designated either for life safety or for utility purposes. When ropes are intended for life safety use, they must be made so that there is a continuous filament fiber within the rope. This is to ensure that the rope will not separate under its working load. Utility ropes are not designed for life safety use and should never be used to support a person. Utility ropes are meant to be used only to haul equipment, as tag lines, or to carry loads other than people. Life safety rope is not only intended to carry the weight of the rescuers, but also the weight of any victims. For that reason, life safety rope is manufactured with a higher degree of reliability and safety. NFPA Standard 1983 defines life safety rope and dedicates its use to supporting people regardless of the application (rescue, fire fighting, emergency operations, or training). A life safety rope may be used only for people. A utility rope is never used for people, and a life safety rope used to support any other load should be taken out of service. At an emergency, it is easy for life safety ropes and utility ropes to be intermingled. Avoid such a situation by making it obvious which rope is for which purpose by color coding, by using approved markings, or by other acceptable means (not paint or other chemicals that can degrade the rope). When using rope, consider the need to use two ropes to secure people. One line is the main line or working line, as shown in Figure 9-13, and the second rope is the safety line in case of failure or damage to the main line or its hardware.

When you are using life safety ropes, you will also use rope hardware devices to change direction and control the movement of the rope. Just as life safety rope must meet NFPA Standard 1983, so must the hardware. Ascent devices, descent devices, pulleys, carabiners, and other equipment used to protect, connect, and control rope (as shown in Figure 9-14) are examples of

FIGURE 9-13 Using this rope and harness, could you retrieve this rescuer if he became injured or entangled?

FIGURE 9-14 A carabiner used to connect a figure 8 descender to a sling.

rope hardware. You will use this equipment to attach the rope to slings at anchor points, to attach the rope to other rescue equipment such as a Sked™ (shown in Figure 9-15), to attach the rope to harnesses (as shown in Figure 9-16), and to construct systems that provide mechanical advantage when using the rope. You can control the speed of a descent by rigging a descending device and controlling the rope with it. Pulleys allow you to change the direction of the pull on

Rescue 145

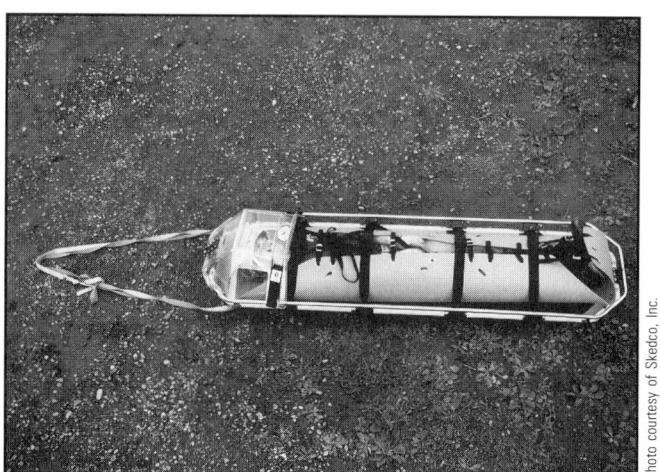

FIGURE 9-15 A Sked™ stretcher placed in a stokes basket.

FIGURE 9-17 A pulley being used to change the direction of pull on the rope by 90 degrees.

FIGURE 9-16 A harness, SAR, PPE, ropes, and other rope equipment being used during confined space rescue training.

the rope, and other devices are intended to be used to protect the life safety rope.

Using rope hardware to create mechanical advantage can be as simple as the change-of-direction pulley shown in Figure 9-17, or include more complicated systems such as two to one or three to one mechanical advantage systems. If you choose to use rope for confined rescue purposes, you must identify your needs, obtain the needed equipment, and train with it. That training must include not only initial training to develop proficiency, but also ongoing training to maintain that proficiency. Maintaining your level of proficiency is not an easy task. It takes commitment of people, time, and money. Ropes used in training should not be used for rescue purposes, and all life safety ropes require inspection and maintenance. Before you purchase equipment and begin training for confined-space rescue, clearly identify what you expect to accomplish.

Utility ropes need to be strong enough hold the load of any equipment you may need to raise or lower, easy to tie, and supple enough that they will hold knots. Though it may seem to be of less concern with utility ropes, watch for sharp edges, rough spots, or contaminants that can damage or cut the utility rope. Utility ropes are often subject to damage and/or overloading because they are not considered as critical as life safety ropes, but you do not want to drop critical equipment when you are hauling it. Losing a fan by having it crash to the ground or dropping a tool into a confined space may change conditions enough that your entire rescue operation fails. To rephrase what was stated previously about life safety ropes and utility ropes: *Utility ropes are never used as life safety ropes, and life safety ropes are never used as utility ropes.* Any life safety rope used for utility purposes should be taken out of service and destroyed.

> **NOTE**
>
> Utility ropes are not meant to be used for life safety and therefore need to be just strong enough to haul any equipment that may need to be raised or lowered.

You can use other equipment for confined space rescue work, as shown in Figures 9-18 and 9-19. Rope yokes designed for use with Class III harnesses that have shoulder-mounted D-rings, wristlets, specialized rescue stretchers such as the Sked™, backboards, low voltage lights or other lighting equipment that will not be an ignition source in certain environments, and communications equipment are just a few of the types of equipment available. The wristlets are designed to go around either the rescuer's or the victim's wrist and then be attached to a retrieval line at the center. This allows the person to be raised or lowered with their arms extended over the head. The wristlets may not be comfortable, but they may be the most effective way to move a person. At a minimum, your confined space equipment must include a rescue stretcher specifically designed to fit through the opening

FIGURE 9-19 A rescuer wearing a supplied air respirator with a radio adapted for use during confined space entry.

FIGURE 9-18 Equipment for confined space rescue laid out in a staging area to organize and account for the equipment.

of the space with the victim in it. Rigid stretchers such as Stokes baskets cannot be manipulated to create a smaller profile and may not pass through the opening with or without a victim in the Stokes basket. If you need to use a backboard to immobilize the victim, make sure it will work effectively with your rescue stretchers.

Electrical Equipment

Electrically powered lighting equipment can create an ignition hazard, and you should consider if you need to purchase listed electrical lighting equipment. Ideally, you want to eliminate the flammable atmosphere, but what if your confined space is configured in a way that it may contain dead air spaces that are difficult to vent? Eliminating the ignition source will provide an additional layer of protection, and the more layers you build in, the better chance you have of preventing a problem. When using electrically powered equipment for confined spaces, do not ignore potential electrocution hazards to your personnel or victim. If you are working on a metal platform or entering a steel tank that becomes energized by a damaged power cord or tool, the electrical hazard may not be apparent until someone receives a shock. Ground fault circuit interrupters (GFCIs) are intended to prevent electrocution. Do you use power cords or electrical sources with GFCI devices? Another answer to the electrocution hazard may be to use low voltage lights and equipment.

Communications Equipment

Confined space rescue requires communication with the rescue team. Visual communication can be effective, but it requires two things. First you must be able to see the rescue team at all times and second, someone must be watching the rescue team at all times. Two-way voice communication is more effective, but voice communication when wearing protective equipment can be difficult. Depending on the age and type of the SCBA or SAR you are using, there may be built-in or accessory voice communication. If you have this equipment, or plan on purchasing the communication equipment, ensure that you understand any limitations of the equipment (range of signal, connectivity with your radios, etc.), its compatibility with your respiratory protective equipment, listings or approvals for the equipment (NFPA standards, etc.) and whether using equipment from different manufacturers (SCBA and communication equipment, that is) will void warranties or affect the safe use of the SCBA or SAR. Get the answers before you buy the equipment and have it demonstrated to you. If the equipment is voice activated, see if breathing sounds are enough to open the microphone. If there is a talk to press button or switch, see if it is in a good position or if it is difficult to use while wearing typical rescue equipment. If you have to remove your gloves to press the switch, is that a good thing? The goal is to improve communications, and equipment that will not work easily while wearing protective clothing will limit your ability to communicate. You might be able to combine voice and visual communications. It might be possible to send a verbal message from outside the space that can easily be answered by visual signals from the rescue entrants. In that case, either keep the answers simple (yes or no), or have a preplanned set of signals that you have trained with and everyone is familiar with. Once again, the ideal is to preplan your communication needs. At the very least, have a standardized means of communication that will allow you to easily apply it to confined space rescue and realize that communication is a strategic factor that you must include in your size-up.

Training of Personnel

The time to begin training is long before you have an incident. Whatever equipment you have purchased, your people must know how to use it properly. This book is based on the requirements of NFPA® 1006 *Standard for Technical Rescuer Professional Qualifications, 2008 Edition*. You should also be aware of **NFPA 1670 *Standard on Operations and Training for Technical Search and Rescue Incidents, 2004 Edition***. This standard divides technical rescue into different areas such as confined space rescue, vehicle and machinery rescue, and rope rescue. Each section in the standard includes general requirements as well as specific requirements based on the type of technical rescue. These requirements include prerequisites that must be completed for awareness, operations, and technician levels. Additionally, there are team requirements including one in section 7.4.2 that mandates a confined space rescue team operating at the technician level consist of at least six people trained to the technician level.

This NFPA standard should be the minimum that you use to identify your training needs. As with any other recognized standard, you can exceed the requirements based on local needs, and you must consider retraining on a periodic basis. Retraining is maintenance to ensure you keep up the necessary proficiency. Just like you would not drive a car with bad brakes and bald tires down a wet mountain road, you should not rely on a confined space rescue team that no longer knows how to use its equipment or how to tie knots correctly.

> **NOTE**
> You may have the best equipment available, but if your people do not know how to use it properly, it is useless.

Re-creating actual emergency conditions during training exercises is difficult. During training, conditions are controlled to minimize risk to personnel. You can generate only so much realism before you endanger the trainees. In addition, you can stop the training at any point, review what is happening, and then correct it. But that is not to say that real emergencies do not provide training opportunities. You should critique all incident operations as part of the termination activities. Honestly, looking at your team's performance to identify both what went right and what went wrong builds a better team. Evaluate the performance of your equipment. Did it operate as it was designed to, or was there a problem with the equipment? Were you able to use it as you trained and planned to?

Termination

To close out the incident, you should take action beyond critiquing the incident. Incident documentation should be completed as soon as possible. Not only does this provide you a written record of the emergency, but it can be used for future training.

You must inspect, test, and maintain the equipment. Simple things such as refilling air cylinders and packing equipment away might be part of your standard routine, but you could have equipment that needs to be sent back to the manufacturer so that they can inspect it and re-certify it. If you have contaminated equipment, can it be decontaminated, or does it need to be properly disposed of? Life safety ropes must be inspected by a qualified person, who must follow the manufacturer's method for inspecting them. Every piece of equipment you have used must be put back into service or replaced so that it is available during the next incident and so that it will perform as intended.

As your team develops, you will also develop personal experience that can be used to prepare for the next incident. Until you develop that store of information, talk to other teams who have had incidents. What did they learn? What were they able to accomplish? Where did they see limitations with their equipment, their operations, or their training? Ask them how and where they were successful and use their experience as a training tool.

INITIAL SCENE OPERATIONS

As you arrive on the scene of a confined space accident, you will want to start operations as quickly as possible. You now have nine steps to initiate your operation. Begin working on each step. Some of the steps fall in line and can be completed in sequence; perhaps others can be performed simultaneously. The key is to identify the problem and establish your goals and objectives. Look to first execute those that have the greatest impact on the situation and are easy to complete. Second, total up the real number of victims. Did others enter the space as would-be rescuers and then become victims? Third, initiate atmospheric monitoring as soon as possible and determine if you face any identifiable atmospheric hazards. Consider ventilation if ventilation is an effectual initial action and determine what value it provides for your victims and your rescuers. Atmospheric monitoring and ventilation can both begin with little or no PPE. Finally, determine if the victim is alive or dead. Look at your incident and the probability that the victim is dead. A victim with traumatic injuries that include decapitation is dead. A victim with exposure to highly toxic chemicals for 10 or 15 minutes may be dead or may die before you can get to her. You need to think of both the victim and the rescuers at this point. If the victim is known to be deceased, you are not performing a rescue, but a body recovery. What you do next and the objectives you choose should reflect that fact. If, however, you feel the victim is alive or are unsure of whether the victim is alive, your goal will be rescue and your objectives and actions will drive you toward getting to the victim and removing her. Whatever you decide to do, remember that when little is to be achieved, little should be risked.

Operations during a confined space emergency can be compared to operations during hazardous materials emergencies. There are two basic types of operations: **defensive** and **offensive**. Defensive operations limit exposure to the hazards by having personnel and others not enter the hazardous area and not come into direct contact with the hazard. In addition, defensive operations can often be performed with little or no PPE, as shown in Figure 9-20. Defensive operations are both life safety– and incident stabilization–directed and can gain time to allow offensive operations to begin. Though most confined space emergencies require offensive operations, don't overlook defensive operations. If it is possible to immediately remove the victim by using the in place retrieval equipment while you are outside the space, do it. When you remove the victim from the space, most of your other problems become secondary or go away.

Offensive operations are more difficult. When you must choose offensive operations, your equipment and resource needs increase, and the actions take more time to accomplish, as shown in Figure 9-21. All of the necessary resources have to be in place before you start an offensive operation. You can't put rescue entrants into the confined space before you have a means to get them out. You can't send in the entry team before you have a backup team available to rescue them if the need arises. Getting the needed rescuers to the scene, briefing them on the action plan, and setting everything up takes time. Does the victim have that much time? Once again, look at the high-impact, easy-to-do actions and see if they can buy time for the victim. Though defensive actions can be directed at both life safety and incident stabilization, offensive operations focus more on life safety. Typically that life safety focus is on the rescue team. The rescuers cannot enter the space without PPE and other specialized equipment, the PPE takes time to put on, the specialized equipment takes time to make functional, and the rescuers need to be briefed on what they are to do. Look to provide a coordinated approach. Identify and complete both defensive and offensive actions. Look to ensure victim survivability until he can be rescued.

Defensive Operations

Performed without entering space or prior to entering space Examples:

Establishing ventilation

Use of monitoring instruments while outside the space

Non-entry rescue or retrieval of victim

Advantages:

Limits risk to rescuers

May assist in stabilizing the incident prior to entry

May be the fastest and most successful way of rescuing victim

Level of knowledge and skill possessed by rescuers at lower levels

Disadvantages:

May not be effective means of rescue of victim

FIGURE 9-20 Both defensive and offensive operations can be taken during a confined space rescue. This chart characterizes defensive operations.

Offensive Operations

Performed by entering the confined space for rescue Examples:

Use of monitoring instruments while inside the confined space

Packaging and removing victim from within space

Assisting another rescuer within the confined space

Advantages:

May be the only way to rescue the victim from the space

Disadvantages:

Requires higher level of risk to be managed

Requires higher levels of training and skill for rescuers

Requires additional equipment for entry, retrieval, and personal protection

FIGURE 9-21 Offensive operations place greater demands on rescuers and equipment.

ACCESSING THE VICTIM

To meet NFPA 1670, each member of the rescue team must be certified to DOT First Responder, have CPR training, and perform a primary survey for any victims in the space (airway, breathing, and circulation). A victim with a compromised airway cannot get air. If your victim is laying face down in a puddle of water, the airway is not maintained. Given enough time, no airway means death. When your victim is breathing or you are able to re-establish their airway and they begin breathing on their own, you have solved that problem. But what about a victim who is not breathing? How will you provide rescue breathing while wearing respiratory protection in a contaminated atmosphere?

As this book is being written, CPR methods are changing and rescue breathing may become a thing of the past for first responders. However CPR may change, work to get a rescuer into the confined space as soon as possible. Once the rescuer is in the space, he can provide a self-contained source of oxygen for the victim. You should consider whether it is practical to change the atmosphere in the space as quickly as possible. The ultimate goal is to remove the victim from the contaminated atmosphere as quickly as possible, as shown in Figure 9-22. Performing compressions during CPR in a confined space may be difficult if not impossible. Again, getting the victim out of the space as soon as possible resolves that problem, but it may present the issue of how to handle a spinal injury. If your victim dies from the atmosphere in the space, the other injuries will not matter. When your victim is viable (airway, breathing and circulation appear good), look for injuries and prepare your patient for transport. Above all, follow local protocols for treating victims. This book is not intended to provide comprehensive information on the treatment of victims.

FIGURE 9-22 How rapidly your victim must be removed from the confined space should be one of the primary considerations in packaging your victim.

VICTIM STABILIZATION

How was your victim injured? Did he fall? Was there a preexisting medical condition? Or was he exposed to a hazard within the space that caused him to fall or pass out? Evaluate the space and work from the facts that you know or develop. Yes, a fall can produce trauma injuries, and you should treat victims as trauma patients. But what caused the fall? Was it some hazard associated with the confined space or did the victim, as the witnesses tell you, simply fall due to something unrelated to the confined space. Initiate monitoring. Despite the precautions rescuers have to take to protect themselves, you have to consider if the victim needs to be immobilized. If so, your rescue stretcher must be able accommodate a backboard, and you must know how to use the equipment.

VICTIM REMOVAL

Remove the victim *in the manner that is best for the victim*. Do not let the removal process become more hazardous than letting the victim remain in the space as you work to create a safer way out. Removing a victim horizontally, as

FIGURE 9-23 The location of victims and your access to them will affect how you can remove them.

FIGURE 9-24 The narrow width of these stairs, combined with obstructions such as this pipe flange, may make it impossible to carry a victim to the ground.

FIGURE 9-25 The location of this manhole, on a steep hillside, combined with the narrow opening will require you to remove the victim in a vertical position.

shown in Figure 9-23, is often better than lifting vertically. If your only available removal method is vertical, ensure your equipment is designed to do the job. Rescue stretchers in the vertical position may make your victim more apprehensive and may present problems with the use of a backboard, and you face the risk of the victim falling out of the stretcher (with or without equipment failure). If you must place the stretcher vertically, as shown in Figures 9-24 and 9-25, make sure that the victim is properly positioned and secured to the stretcher. Double-check, prior to the lift, that you have fastened the victim to the stretcher as needed for a vertical lift. Do not attempt to use a stretcher that is not designed to be used in the vertical position. Once again, training is the key to your success. The degree of competency needed to package and lift a patient must be 100 percent. Load test the stretcher and equipment before you begin to raise the stretcher out of the space. Bringing the stretcher just off the ground and allowing the full weight of the victim and stretcher to load the equipment is one way to do that. Don't forget to check fasteners, knots, straps, and the like to be sure they are in the proper position and are holding the victim in place. A mis-positioned strap or knot may be uncomfortable for the victim, but it may also cause additional injury. Only after you are satisfied that everything checks out okay do you begin the lift. If conditions are not satisfactory, take the load off the stretcher and fix the problem. As stated earlier, your level of training and competency are invaluable. You will not have a procedural checklist in the space with you, and you must rely solely on your training.

LESSONS LEARNED REVISITED

You are now faced with a 30-foot deep manhole with 15 feet of water in it, and there is a partial collapse of the wall where the plug was in the pipe. There is a lot of information flying around at this point. The work crew consisted of 5 people. You now have the injured foreman, the injured worker, and two workers administering first aid. Where is the fifth worker? The answer to that question and a full head count of the members of the work crew will direct everything you do from this point on. If the fifth worker is missing, where do you start? If people tell you that the fifth worker has been taken to the hospital already, how do you ensure that this information is correct and this person has really been accounted for?

Discussion Questions

1. If you need to call additional resources, where will they come from and how long will it take to get them there?
2. What actions do you want to take at this time?
3. What actions will you take when your additional resources arrive?

SUMMARY

- The nine-step process for confined space rescue is:
  - Establish command and take control of the incident.
  - Identify the type of rescue problem.
  - Perform hazard and risk assessment.
  - Identify rescue objectives.
  - Identify resource needs.
  - Develop an action plan.
  - Implement the action plan.
  - Evaluate the effectiveness of the action plan.
  - Terminate the incident.
- Initial operations should be those that have the highest impact on the incident and are the easiest to initiate.
- Work to get to the victim as soon as you can while ensuring your own safety. When you get to the victim, take steps to stabilize her and address her medical condition. Remove the victim as appropriate for her injuries and medical condition.

REVIEW QUESTIONS

1. Identify and briefly explain each of the nine steps for confined space rescue.
2. Explain how tripods should be loaded for lowering or raising a rescuer or victim. If the load cannot be applied in the proper direction, explain how it can be transferred so that it is applied in the proper direction. You may use a diagram to explain, but you must show the direction in which the load is being applied.
3. What NFPA standard addresses life safety rope?
4. Explain the difference between defensive and offensive operations and the value of each type at a confined space emergency.
5. Tripods for confined space rescue can be used to raise and lower both people and equipment. True or False?
6. NFPA 1670 requires a confined space rescue team operating at the technician level to consist of at least how many people?
   a. 2
   b. 4
   c. 6
   d. More than 6
7. The best rescue equipment available is only as good as the people trained to use it. True or False?

KEY TERMS

critical incident stress 136
tripod 137
laid ropes 143
braided ropes 143
kernmantle ropes 144

Standard 1670—Standard on Operations and Training for Technical Rescue 147
defensive actions 159
offensive actions 159

ADDITIONAL RESOURCE

For additional information on the incident referenced in the Lessons Learned section, you can view the FACE report at the following link: http://www.cdc.gov/niosh/face/stateface/mi/07mi024.html

10 Standard Operating Procedures

LESSONS LEARNED: TWO UTILITY WORKERS INJURED BY A STEAM LINE

Two employees working in an underground vault sustained serious injuries when a steam line they were repairing was re-energized. The two workers were part of a three person work crew that was sent to repair a leaking flange located in the vault. The flange was in a steam line that interconnected two steam plants.

Earlier in the day, the crew had the line shut down by the plant operator in steam plant 1, but they did not lockout/tagout the valves controlling the steam line. The crew had difficulty repairing the flange and took longer than expected. After returning from lunch, two members of the crew re-entered the vault and were working in the vault when the operator from plant 1, believing that the work had been completed, reopened the valve.

Critical Thinking Questions

1. Do you believe that written work procedures would have prevented this accident?
2. What one procedure, had it been followed, could have prevented this accident?
3. Can you identify any other procedures that you feel would be necessary for this type of work?

LEARNING OBJECTIVES

NFPA 1006, *2008 EDITION*
By the end of this chapter, you should be able to:

- Use a standard approach for confined space rescue

INTRODUCTION

Every day of our lives we use standardized methods of doing things. The route we take back and forth to work or to the homes of our family and friends are near identical every time we drive them. But have you ever had to give someone directions so that they could follow the same route? As you tell them how to go, you know all the turns, but probably can't remember every street name. It's not that you don't know how to get there, it's that now having all the information and getting it correct so that someone else can do the same thing you do matters most of all.

DEVELOPING STANDARD OPERATING PROCEDURES

Have you ever been assigned to set up a gasoline powered rescue tool during a drill and you have everything connected, but you can't start the engine? Someone else comes and tells you that you need to choke the engine when it is cold. In all the training you have had, you've seen the engine started different ways and you have never been shown the correct choke position. During the drill, the tool will eventually get started even though it took longer than expected, but it was only a drill. What would happen, though, if you had to use that tool at an emergency? Would you know the correct setup and startup procedure, from start to finish? That is what **standard operating procedures (SOPs)** do for you. They give you a proven method you can use to accomplish a particular task or objective. SOPs are based on an effective process or technique that is repeatable. That means if you follow the procedure every time, you achieve the same results every time. If you are monitoring the atmosphere in a confined space or setting up a tripod, don't you want to make sure you have tested the entire space or set up the tripod so that it will be stable? Procedures are not unique to confined space rescue. We use them every day. The routine you follow when you get into your car to drive someplace is an example of a daily procedure. If you forget a simple step such as repositioning your seat because someone else drove the car before you, what are the consequences? You stop and take a minute or two to adjust everything. When you go to an ATM to withdraw cash, you also follow a procedure. The ATM prompts you through the procedure with instructions on the screen. In the event you miss one step, or do not complete it correctly, you don't get any cash out of the machine. Why the difference between driving your car and getting money from an ATM? The answer is in the consequences of that missed step. In the case of your car, the seat in the wrong position may make it uncomfortable to drive, but it will probably not cause an accident. With the ATM, a loose procedure can allow unauthorized people access to your accounts, and they can take your money. When the price of failure can be measured in safety and reliability, you must develop written procedures to manage those risks. The SOPs you develop will be written processes based on experience, knowledge, and understanding of what you are attempting to do. SOPs will not, typically, be the exact, unwavering method to achieve an objective. There may be some flexibility in the SOPs because they will be based on an anticipated set of circumstances that you may have before you. Some situations will not fit perfectly with the procedure, but having a procedure, understanding the basis for it, and understanding how it is meant to be applied will allow you to relate the procedure, within reason, to the circumstances you are facing.

Not only must SOPs be accurate, they must also be current. Failing to recognize changes that take place over time can render an SOP useless. A firefighting SOP that requires a ground ladder to be thrown to a second-floor, street-side window for every working fire in residential buildings is of value when your buildings are two stories. The intent of such an SOP is to provide a reliable second path of escape for any firefighters caught above the fire. They know if they are cut off by the fire, getting to that ladder is their way out of the building. The SOP might even have originated from an earlier incident during which firefighters were cut off and had no way out. The problem arises when the community changes. The neighborhood is no longer residential, but now consists of five-plus-story office and commercial buildings. Yet firefighters continue to raise a ladder to a second-floor window of every building regardless of how tall it is. Not only is this SOP outdated, it is a waste of people. The time and personnel used throwing the ladder at a five- or ten-story building is wasted. The people and time can be used elsewhere. The firefighters throwing the ladder have no idea as to why they're throwing the ladder except that the SOP says they have to.

To get the most value out of having an SOP, people must understand the background and purpose of the SOP, they must be trained in its application, and it must be updated. Updating should begin as soon as the SOP is written. While you are developing the SOP, take it out into the field and walk through it. Let the people who will have to use it try it out. Have them critique the SOP and ask questions. Once the SOP has been developed, train the users. Teach them to recognize the limits of the SOP and give them enough knowledge to decide when to apply it and when to deviate from it. Personnel should not be taught to blindly follow a procedure. If the user must follow the steps in order and must complete every step (such as donning an SCBA), make sure they know that and are aware of the consequences of not following the steps in order.

WRITTEN SOPS

Written SOPs are formal documents that should be in a standard format. You record the information so that it can be passed from person to person, with identical information being passed each time. When the significance of following the SOP step-by-step or in the same way each time increases, the more important the written SOP becomes. Developing a written confined space entry program is an example of just such an SOP. The intent of the entry program is to allow people to safely enter and work in confined spaces by reviewing and measuring potential conditions in the space against what is known to be safe or unsafe. Only after those steps are completed and the hazards are either not found or are identified and controlled, will entry be allowed. This is a simple look at an entry program; an actual program is more complex than that, but you can see how knowledge and accuracy require written SOPs.

When you begin looking at developing an SOP, look at those things that are similar between incidents. You may think that every incident is different, but there will be some common threads running through them. Confined space incidents share the same features that define the location as a confined space. SOPs for equipment, such as one for setting up a tripod, will require you to do certain things the same way every time. How you conduct atmospheric testing or identify the need for lockout/tagout may vary slightly from incident to incident, but each task has repeatable steps that allow for an SOP to be created. That similarity allows you to identify and address expected problems, and it does not require you to reinvent the wheel each time. It also helps to ensure that you do not miss essential steps. SOPs can be detailed enough that they provide a set solution to a specific problem (for example, testing of SCBA), or they can be general and give you basic guidance for a situation (for example, establishing command).

Chapter 9 addressed the nine-step process for confined space rescue. Those nine steps should be where you begin to develop your basic SOPs. You must have a method to initiate your operations that is reliable and repeatable so that you will cover the basic elements common to confined space emergencies. Once you have taken care of the basics, you can expand your operations from there.

Keep the basis for your SOPs simple, but do not make your SOPs so simple that you find them inadequate during an emergency. Fill in details that need to be addressed. If you develop an SOP to establish command, define it to the point that everyone knows how command is expected to be established, how it will be passed, and how the IMS will be expanded to meet the needs of the incident. If you have determined that one of the first command staff positions to be addressed after command has been established is the safety officer position, specify that in your procedure. You should also go a step further and describe the type of person who would be a competent safety officer and what authority they have. Assigning responsibility without authority will accomplish nothing. Does the safety officer have general incident safety responsibilities, or are they assigned specifically to the confined space rescue operation? Who do they report to, and do all the other emergency workers know what the safety officer has the power to do?

The SOP is a record of what is to be done to successfully complete a specific task. When the SOP was written, there were certain conditions anticipated, and the assumption is that following the SOP will allow you to meet a certain level of performance. SOPs are written for routine items and, at times, for unique situations in which you have accumulated a certain amount of data and can reasonably expect what will occur. This is why you can use each of the nine steps for confined space operations and develop basic SOPs that meet your needs. Pre-planning provides solid information for emergency response. It can be checked for accuracy long before any emergency; when you are faced with the real incident, you will have to compare it against what you actually have available in equipment, personnel, resources, and the real-time rescue conditions you are faced with. The SOP serves as a reminder of what you must consider and the action you must take at each confined space incident. If the rescue situation does not fit within your existing SOP, do not abandon the SOP. There will be some similarities. At least consider what is usable in your existing one and consider if it provides alternative methods to meet the challenges with which you are faced. You may even be able to develop your SOPs to the point that you anticipate different conditions and identify alternatives in the SOP. Realizing that the incident you are faced with exceeds your abilities and having pre-arranged available assistance is just such an SOP.

Though SOPs may be meant to serve as basic guidelines, they should be developed to the point that they are thorough. SOPs should be the basis for training and drills in advance of an emergency; the first time you read them should not be as you arrive at the emergency. Using SOPs for training means that everyone gets an idea of how the operation should begin and what roles they play in the rescue. Applying SOPs during training also allows you to evaluate the SOPs and revise them as needed. Having written SOPs also means that you can go back to the SOP after the training and have access to that exact same information in the future. It prevents the loss of information that occurs when the SOP is only verbally transmitted from person to person. When the life of a victim or the lives of emergency responders depend on 100 percent accuracy, do you want the person who is in charge to know what needs to be done, or will you accept that they might be half right? Leaving crucial steps in the setup of a piece of equipment to memory can lead to disaster. Writing an SOP and then creating a checklist based on that SOP works to guarantee accuracy and success.

In your SOPs, clearly identify the type of situation they are to be applied to and to whom they apply. If different levels of training are required for different roles, then the SOP should have a section that addresses those differences in training and roles. People have different abilities to perform assigned tasks based on what we want them to do and how well we train them to complete those tasks. In technical rescue operations, just as in hazardous materials response, there are different levels of training (awareness, operations, technician) and different types of actions (defensive or offensive) that can be taken based on that training. The first people to respond to the confined space emergency may only have awareness training. They might just get there and recognize the problem. Or you might have two qualified rescue technicians discover and report the incident. In either case, there are limitations to what they can do. An SOP that initiates the response to the incident and gives direction on what to do until the full rescue team is on the site should be considered, and it should be different than the SOP for the full team. Regardless of how you decide to write the SOPs, write only those that you need. The SOP is a guideline built on the methods that were successful in accomplishing incident objectives for you or for others who have emergency response experience and knowledge. No SOP can address every situation. Rely on the similarities between incidents to find a common thread to use. That commonality can be a strong, integral part of your operation. When that common thread is weak or nonexistent, your SOPs may be nothing more than a starting point to allow you to create a response to the incident.

CHECKLISTS

Standard operating procedures are great tools that allow you to formalize your operations in order to ensure success. But how do you make practical use of an SOP in the field? Using the SOP in training allows you to become familiar with it, but it does not mean you have memorized it. Applying an SOP and making sure that you thoroughly cover each critical step in the proper sequence is difficult. Having a multipage SOP in the field does not necessarily mean it will be any easier. Develop checklists for your SOPs that condense the SOP into a logical progression of what needs to be done, as shown in Figure 10-1. A checklist coincides with the SOP, briefly identifies critical points, and allows you to record the completion of each step. Checklists also allow you to identify and record when the steps do not apply in the particular conditions you have. Using a one- or two-page checklist is much simpler than trying to read the five- or ten-page SOP. You can also use checklists as an evaluation tool during training sessions. Use the checklists to identify when the trainee completes each required step successfully. If the required steps are not completed or the student is having difficulty completing a step, you may be able to identify and record areas in which additional training may be necessary. If the failure is widespread, you may also have identified limitations in the capabilities of your personnel, equipment, SOP, or training methods. If your SOP is based on rescuers making entry wearing SCBA, but during training you see that the students are having difficulty donning the SCBA, where is the problem? Is the problem with your SOP, or with the SCBA qualifications of the students? When you identify problems, look to fix them. If your confined space rescue team is qualified in every area but they do not know how to wear SCBA, they aren't a rescue team and do not belong in confined spaces. Make the changes where necessary and adjust your operations and SOPs to reflect those changes.

Step 1—Establish command

Command established and identified	☐ YES	☐ NO

Command is: _____
Command has been transferred to: _____

Step 2—Identify the type of rescue problem (choose only one)

Confined-space rescue	☐ YES	☐ NO
Rope rescue	☐ YES	☐ NO
Hazardous materials incident with rescue	☐ YES	☐ NO
Other nonfire rescue	☐ YES	☐ NO

Step 3—Perform hazard and risk assessment

Confined-space entry permit present	☐ YES	☐ NO	☐ N/A
If "NO" can processes and/or potential hazards be identified?	☐ YES	☐ NO	☐ N/A

Date and Time Issued: _____ Date and Time Expires: _____

Job site/space ID correct	☐ YES	☐ NO	☐ N/A
Job supervisor identified and present	☐ YES	☐ NO	☐ N/A
Equipment to be worked on identified	☐ YES	☐ NO	☐ N/A
Work being performed identified	☐ YES	☐ NO	☐ N/A

List: _____

All on-site personnel accounted for (include victims)	☐ YES	☐ NO

Number of victims: _____ Number of on-site personnel _____
Location of victims: _____

Required MSDSs at site and available	☐ YES	☐ NO	☐ N/A

Atmospheric Checks: Check once before ventilation, continuously monitor, and then log at 15-minute intervals after ventilation.
Oxygen: <19.5% oxygen deficient: > 23.5% oxygen enriched, Explosive: >10% LFL potentially explosive, Toxic see TLV or TWA.

Time	Oxygen—%	Explosive %LFL	Toxic—PPM

Identity of Toxic Materials _____

FIGURE 10-1 A confined-space rescue checklist.

Step 4—Identify rescue objectives

Victim rescue:	☐ YES	☐ NO
Can you identify defensive operations?	☐ YES	☐ NO
Can you identify offensive operations?	☐ YES	☐ NO
Victim recovery	☐ YES	☐ NO

Record your objectives:
_____
_____
_____

Step 5—Identify resource needs to support rescue objectives

_____
_____
_____

Direct reading gas monitor—tested	☐ YES	☐ NO	☐ N/A
Safety harnesses and lifelines for entry and standby persons	☐ YES	☐ NO	☐ N/A
Hoisting equipment	☐ YES	☐ NO	☐ N/A
Powered communications	☐ YES	☐ NO	☐ N/A
SCBAs or SARs for entry and standby persons	☐ YES	☐ NO	☐ N/A
Protective clothing	☐ YES	☐ NO	☐ N/A
All electric equipment low voltage and intrinsically safe and nonsparking tools	☐ YES	☐ NO	☐ N/A

Step 6—Develop an action plan

_____
_____
_____

Ventilation

Ventilation started:	☐ YES	☐ NO	☐ N/A
Positive pressure mechanical	☐ YES	☐ NO	☐ N/A
Natural ventilation only	☐ YES	☐ NO	☐ N/A

Communication procedures: _____

Rescue Procedures

Rescue procedures identified and communicated to rescue team for:			
Victim(s)	☐ YES	☐ NO	☐ N/A
Rescuers	☐ YES	☐ NO	☐ N/A

Entry, standby, and backup persons:	☐ YES	☐ NO	☐ N/A
Completed required training?	☐ YES	☐ NO	☐ N/A
Is it current?	☐ YES	☐ NO	☐ N/A

FIGURE 10-1 Continued.

Step 7—Implement the action plan
The following has been reviewed and is acceptable for rescue operations to begin:

Lockout/deenergize/tagout	☐ YES	☐ NO	☐ N/A
Line(s) broken-capped-blanked	☐ YES	☐ NO	☐ N/A
Area secured (post and flag)	☐ YES	☐ NO	☐ N/A
Breathing apparatus—SCBA or SAR	☐ YES	☐ NO	☐ N/A
Full body harness w/D-ring or other acceptable harness	☐ YES	☐ NO	☐ N/A
Emergency escape retrieval equipment set up and checked	☐ YES	☐ NO	☐ N/A
Lifelines	☐ YES	☐ NO	☐ N/A

Respiratory Equipment Time Log

Rescue Team 1		Rescue Team 2	
Name _____	Name _____	Name _____	Name _____
SCBA or SAR (circle)	SCBA or SAR (circle)	SCBA or SAR (circle)	SCBA or SAR (circle)
Pressure _____	Pressure _____	Pressure _____	Pressure _____
On air time _____	On air time _____	On air time _____	On air time _____
Air duration _____	Air duration _____	Air duration _____	Air duration _____
Recall time* _____	Recall time* _____	Recall time* _____	Recall time * _____
Off air time _____	Off air time _____	Off air time _____	Off air time _____

*Recall time is on air time plus air duration time minus a minimum of five minutes for time to exit the space. For longer exit times, subtract more time from the combined air time and air duration time.

Step 8—Evaluate the effectiveness of the action plan

Plan is proceeding as expected	☐ YES	☐ NO
Plan is not proceeding as expected	☐ YES	☐ NO
Minor changes needed?	☐ YES	☐ NO
Major changes needed?	☐ YES	☐ NO
Changes communicated and effective?	☐ YES	☐ NO
Does the operation need to be stopped to make changes?	☐ YES	☐ NO

Changes to action plan: _____

Step 9—Terminate the incident
Check that all personnel are accounted for.
Retrieve, recover, and maintain equipment as needed.
Prepare incident reports.
Hold a debriefing and critique.
Evaluate operations and their effectiveness.
Consider if remedial actions are needed.

FIGURE 10-1 Continued.

LESSONS LEARNED REVISITED

The third employee called for help when he saw steam entering the vault and the plant 1 operator immediately shut down the steam line. The two injured employees were able to climb out of the vault on their own and were transported to a nearby burn center, where they died from their injuries several weeks later.

Discussion Questions

1. Based on your answers to the first three Lessons Learned questions, which procedures would also be applicable to this confined space rescue if your team had to enter the space?
2. If you had responded to this incident and the plant manager told you she had steam protection suits available to use, would you use them at this incident?
3. If your ESO had steam protection suits, or other highly specialized PPE, what criteria would you use to identify the people qualified to use them, and how would you supervise the donning of the suits?

SUMMARY

- Standard operating procedures allow you to formalize your operations to ensure success.
- Identify and cover each critical step in the proper sequence.
- When necessary, develop checklists for your SOPs to condense the SOP so that it is usable in the field, as shown in Figure 10-1.
- Train with your SOPs and checklists.
- Use your SOPs as evaluation tools.
- Regularly evaluate and update you SOPs and checklists.
- When you identify operational problems, fix them.

REVIEW QUESTIONS

1. Using the nine-step process in Chapter 9, identify the minimal SOP that you would develop for response activities at a confined space rescue.
2. Taking the information you developed in question 1, identify those areas that would merit development of expanded SOPs. Choose one area and develop an SOP to address that area.
3. Based on the SOP you developed, create a checklist to support it. Explain how you would use the checklist for training and at an emergency.
4. You have a standard operating procedure for a specific type of emergency response; however, if you follow the SOP, this incident will only get worse and endanger more people. You should _____.
   a. Stick with the SOP to avoid trouble
   b. Modify the SOP so that it fits all emergencies
   c. Call for help so that you can have someone else figure out how to apply the SOP
   d. Accept that this incident is outside the scope of the SOP and develop a plan for handling the incident
5. Checklists briefly identify each critical point in an SOP; therefore, you can effectively use a checklist before you have been trained in the SOP. True or False?

KEY TERM

Standard Operating Procedures (SOPs) 155

ACTIVITIES

1. Identify the common features of confined spaces and the common hazards that can be present at a confined space incident.
2. Take those common features and develop a checklist that will allow you to address each feature and hazard at a confined space incident. Hint: look at a confined space entry permit.
3. Based on your checklist, determine what checklist items would be performed in the same manner for most confined space incidents.
4. Using those repeatable performance items, identify which items can be addressed using an SOP.
5. Write one SOP for an identified performance item.

ADDITIONAL RESOURCE

NFPA 1670 Standard on Operations and Training for Technical Search and Rescue Incidents, 2004 Edition

11 Rescue Equipment

LESSONS LEARNED: WORKER INJURED IN FALL FROM A WORK PLATFORM

A 50-year-old maintenance worker was injured when he fell from a personnel platform that had been lifted 12 feet by a forklift. The platform tilted when the worker leaned out of it to wash a wall, and he fell over the platform's guardrail. Though the platform was designed to be used with a forklift and had a 35-inch guard rail around it, the forks had not been placed into the channels in the bottom of the work platform; it was not secured to the forklift and was only resting on the top of the forks.

Critical Thinking Questions

1. Why was the work platform stable enough to support the worker as he was lifted but became unstable and fell as he leaned out of the platform?
2. If the worker had been wearing a fall protection device, would it have prevented him from falling out of the basket?
3. What type of harness should the worker have worn had the platform been properly secured and fall protection devices used?

LEARNING OBJECTIVES

By the end of this chapter, you should be able to:

- Understand the use of rope and related rescue and retrieval systems
- Describe the operation of rescue and retrieval systems
- Understand the proper use of the selected transfer device

INTRODUCTION

Throughout this book, you have read that all rescue operations contain a certain amount of risk and that you must be a risk manager to protect both victims and rescuers. The discussion has included hazardous atmosphere lockout/tagout, incident management, and many other items. Many emergency responders simply accept that risk is part of the job they perform and though it is, the risk should be weighed against the benefit to be received, and it should not lead to unnecessary exposure. Most of us would not think of working on a patient without using protective gloves or entering the hot zone at a haz-mat incident without PPE. We face known and expected risks, and incidents are often so well managed that we may not be aware of the risk. Periodic hydrostatic testing of compressed gas cylinders (SCBA cylinders and oxygen cylinders), performance standards for tires on response vehicles, testing of fire hose, and specifying biohazard protective equipment to meet expected hazards are all examples of risks that are managed without the risk being obvious. We use our procedures, settle into a routine, are comfortable with what we do, and apply our experience and knowledge without having to consciously think out every step we are going to take. What do we do when an extraordinary incident comes our way? Can we identify the hazards? Do we have sufficient time to control the risks and still successfully rescue the victim?

The key success at a complex incident is to break it down into smaller components. Stick with the basics and use your skill, knowledge, and experience to apply the basic concepts. A large space still requires air monitoring, just in more areas and over a longer course of time. You may have to break the space into zones or areas, monitor the area, and, if is safe enter, work your way through one area to get to the next area. What do you do, though, when you are faced with a more complex situation such as a deep, open-topped pit where the only access to the bottom is by way of a narrow spiral staircase on the outside wall? Again, stick to the basics. Many of the same details run in the background of most emergency operations. Know what your equipment is designed to do and what its limitations are. You can adapt the equipment as long as you stay within those limits, and the equipment remains reliable and will perform satisfactorily. But adapting equipment runs the risk of violating those design limits, which is an unacceptable level of risk. You might get away with using the equipment improperly and still succeed in your rescue operation, but would you knowingly accept that risk to emergency personnel and victims? Don't let the stress you may feel while you attempt to rescue a victim pressure you to take on unnecessary risk. You must manage the incident; don't be reckless and don't work outside the limits of your resources.

Most people think that taking on great risk to save others is heroic—you place the lives of others above your own. Under the right conditions it is heroic, but not when you choose to be reckless. If you are reckless, you can injure or kill others by your actions. With reckless actions, you can create a reputation as a person whose judgment may not be trusted in future situations. If you excuse your reckless actions by saying you didn't know better, how professional are you? Professionals know how to use their equipment, they know the limits of the equipment, and they do not needlessly endanger others. Remember that personnel training and education is a limiting factor. When the level of training is low, your expectations of your personnel's abilities should be lower. Without knowing the basics, how can you address basic rescue situations? Sometimes, even though you know the basics, you will not know how to use a new or different piece of equipment. Knowing the equipment includes the design of the equipment and its intended use. This chapter is designed to help you understand your equipment in general. It is not intended as a substitute for learning your specific equipment.

TYPES OF LOADS

Retrieval equipment is intended to be used to raise and lower rescuers and victims during a confined space emergency. The anticipation is that a load will be applied to the equipment and there will be limits to those loads. You must also be aware that there are different types of loads. These different types of loads create different forces and affect how the load is applied. Initially, the weight of the person or object being supported by the device will be the load. Weight is the force being applied by the pull of gravity, and we usually anticipate this load to be applied vertically. When we apply the effects of acceleration or change direction (horizontal or vertical), conditions change and we can easily overload the equipment.

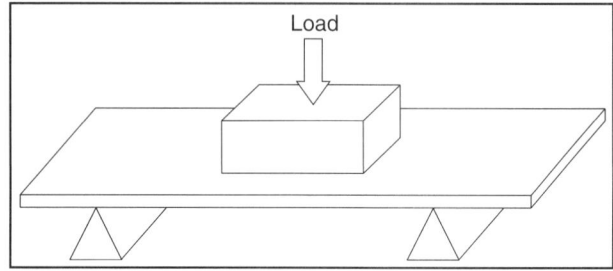

FIGURE 11-1 A static load is applied in only one direction and the whole system is at rest.

The different types of loads that you should be aware of include *static loads*, *impact loads*, *working loads*, *axial loads*, and *eccentric loads*. To understand these loads, you need to understand how each is applied.

Static loads are applied and remain in the same position and location, as shown in Figure 11-1. Normally, the force of the static loads is applied in only one plane (horizontal or vertical). Examples of static loads are forces applied to a harness or life safety rope during testing, a building sitting on its foundation, and a tank sitting on its supports. The object does not move and is not expected to move (unless it fails). For rescue purposes, most loads that you will see in the field are not static loads. Those loads and forces will be moving and may move in several different directions at the same time. As rescuers, we often use static load information when we specify equipment. Equipment is often tested using static loads (among others) to ensure that the piece of equipment meets the requirements of the applicable standards for that type of equipment.

Impact loads are often present in rescue situations and become more critical because they can multiply the force being applied. A load in motion (for example, a rescuer swinging on a rope) will give up some of its energy to another object when it is brought into contact with the object (whether stationary or moving), as shown in Figure 11-2. When the energy is given up in a very short time, due to the acceleration, the applied force will be greater than just the weight of the object. When the energy is applied slowly, the effects will not be as great (think of a hammer swung slowly or quickly). An impact load occurs when one object is in

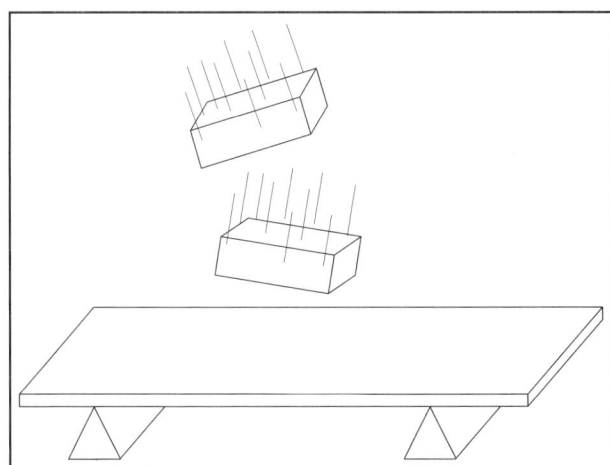

FIGURE 11-2 An impact load is created when a load that is in motion is applied to the support.

motion and comes into contact with another object. The transmitted load will vary based on the forces created by the motion. A person walking on a board is an example of an impact load. When the board is supported on both ends and a person slowly lowers his weight onto the board, it might flex and bend but still support his weight, as shown in Figure 11-3. This impact load is being applied slowly and is within

FIGURE 11-3 During impact loading, the load is in motion, and the acceleration increases the effect of the load.

FIGURE 11-4 The result of an impact load can be great enough to cause the support to fail.

the limits of the board. If you took that same board and jumped on it while it was supported on each end, you would expect to break the board, as shown in Figure 11-4. You don't weigh more when you jump on the board, but the speed with which the force was applied and the acceleration involved greatly multiplied the force of the load. Impact loads can occur frequently during confined space rescues simply because of movement. This movement can be in several directions (up, down, left, right, front, or back) at any one time, and you may be able to control the load, but an impact load may occur unintentionally. The multiplying effect of the impact mandates that equipment used to support human life be able to withstand reasonably expected loads (often measured as weight). To protect against the effect of impact loads, lifelines require a minimum breaking strength of 4,500 foot pounds (lbf).

NOTE

A foot pound is the measure of energy used when a force of one pound is moved a distance of one foot in the direction of the force. Simply using weight as the measure would not take into account the effects of movement.

The **working load** is the expected load applied to equipment during use. The maximum working load is the maximum weight that is expected to be supported by the equipment.

Axial load refers to the direction that the load is carried. *Axial* simply means "moving about the axis," as in the Earth's axis. When a load is carried to the ground, the **axis** is the centerline of the load-bearing column. If you were to look straight down on a tripod from above, imagine a vertical line going straight down through the center of the tripod head. As you lift a victim with the tripod, the direction of pull on the line supporting the victim is straight up and down, as shown in Figure 11-5. As long as the direction of pull is in this position, the load is applied along the axis of the tripod. Individual legs carry that load to the ground equally along the axis of each leg, as shown in Figure 11-6. As long as you load the tripod along its axis, you are using it as it was designed to be used, and you would not expect it to tip or be pulled over. Should you shift the load at an angle to the axis or off center of the axis, as shown in Figure 11-7, the tripod could fail.

Off center loads are called **eccentric loads**. Eccentric loads can cause collapse or twisting of a support. Whenever you set up your equipment, for rescue or training, consider the direction of pull that you will be applying to the load and whether it will affect stability at the point it is applied.

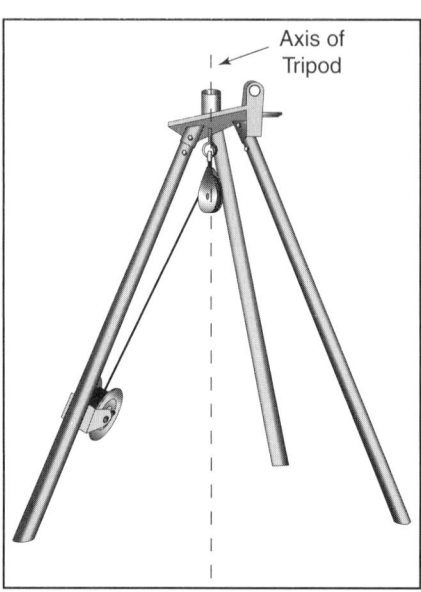

FIGURE 11-6 Even though a tripod has three legs to carry the load to the ground, there is still an axis for the entire tripod. Additionally, each leg of the tripod has an axis.

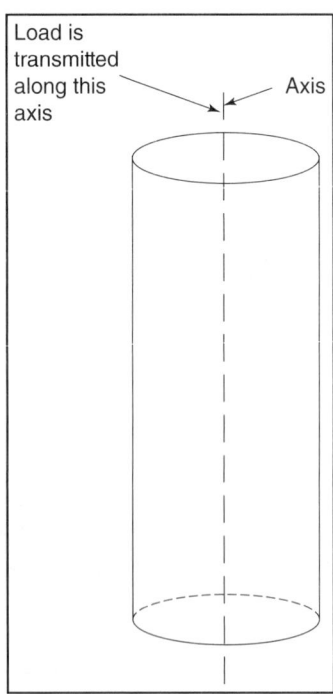

FIGURE 11-5 An axial load is applied in the same plane as the axis of the support.

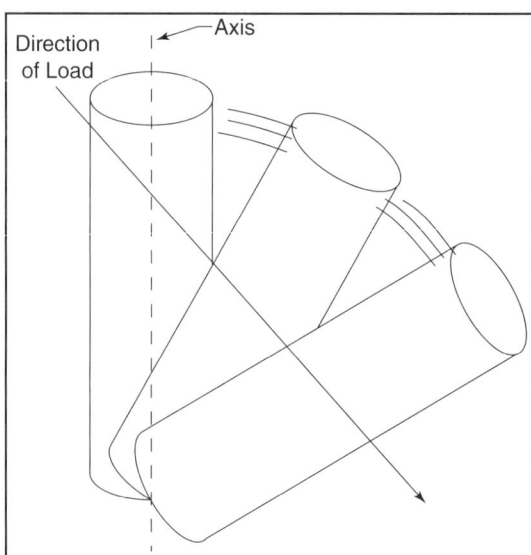

FIGURE 11-7 An eccentric load is one that is applied off center. The eccentric load can cause a failure of the support.

EQUIPMENT STANDARDS

Earlier in the book we discussed the topic of standards. Different standards carry different legal weight. Some standards have been developed and adopted by regulatory agencies, as shown in Figure 11-8, such as OSHA and CALOSHA. These legally adopted standards have the force of law. There are other standards such as those from NFPA and ANSI that are consensus standards and do not have the force of law unless they are specifically adopted or referenced by a government agency. Organizations such as NFPA and ANSI are nationally recognized and have specific procedures regarding the development and adoption of a standard by the organization. Nationally recognized standards may be used in court cases and can be enforced as a result of a court case that creates case law. Whether or not nationally recognized consensus standards have the force of law, they are valuable tools that can assist you in selecting and specifying equipment. Using standards such as NFPA and ANSI means that you will not have to design your own equipment each time you purchase it, and you will provide all of the vendors wanting to sell you equipment with the same standard. To do this you must understand the standards and how they apply to the equipment you are going to purchase.

> **NOTE**
> You may often hear that equipment is OSHA-certified or NFPA-certified, but no equipment is so certified.

OSHA, ANSI, and NFPA do not certify equipment and they typically do not test for compliance with their standards (UL is an exception). Instead the standard is used as the criteria to which a piece of equipment must be designed and manufactured. The manufacturer is responsible for using a recognized third-party testing laboratory to test representative samples of the equipment as it is manufactured (for as long as it is manufactured). The test samples must meet the requirements of the applicable standard. Often the standard requires a specific procedure to be followed for testing the equipment to ensure compliance. By specifying the testing requirements to be used, the test is performed in the same manner each time. This way the test results should be the same each time for identical equipment. When the equipment is tested as defined by the standard and it passes the test, it is then labeled as meeting the particular standard, as shown in Figure 11-9. There are record keeping requirements for both the manufacturer and the testing lab to document that the equipment has passed the requirements. You should request documentation when you purchase equipment.

Meeting standards is expensive for a manufacturer. It is something that the manufacturer can be proud of because it shows their equipment will perform as intended. On rare occasions some manufacturers may try to avoid a particular standard by outright fraud. This can include false labels or referencing another standard from the same standard organization and simply stating that it is "compliant." The NFPA has a specific

FIGURE 11-8 There are a variety of standards that affect how different types of equipment are designed, manufactured, and used.

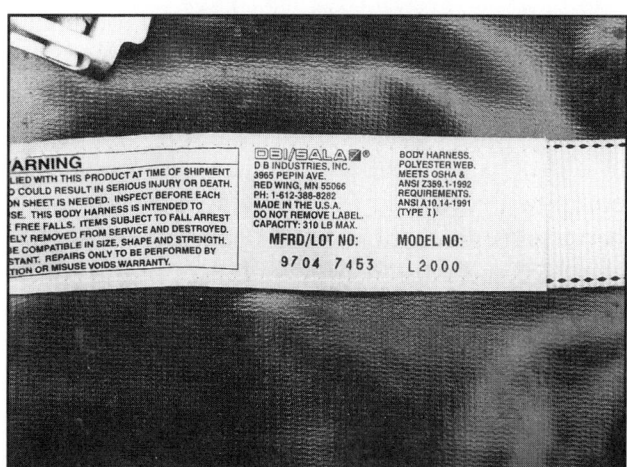

FIGURE 11-9 This label, on a harness, shows the standards that the harness is designed to meet, the manufacturer, lot number, and model number.

standard for Personal Alert Safety Systems (PASS) units—NFPA 1982. When PASS units were first introduced to the fire service, one company was selling a PASS that had not yet passed the testing procedure in compliance with NFPA 1982. This company said that its PASS unit was NFPA-compliant, but it referenced another NFPA standard that was designed to cover electrical equipment in hazardous locations. If the buyer was not aware of the correct NFPA standard, they purchased PASS units that might not operate under firefighting conditions. Fortunately, NFPA was made aware of the problem and the company was forced to change its advertising.

Standards, like any other document, can become outdated, so organizations periodically update them. When you purchase equipment, make sure that you refer to the current version of the standard. How often the standards are updated varies with the standard making organization. If you do not know which version is the latest and greatest, contact the standard organization to find out the date of the most current standard. Older equipment purchased under an earlier version of a standard is generally still usable, but you should check the new version to see what changes have been made. Updates to standards may require different load ratings, or there may have been changes to the testing procedures. Your equipment should be used as designed. If you see newer equipment used in a different way, it does not automatically mean that you can use your existing equipment in the same way.

At times standards from different organizations disagree when they cover the same or similar topics. Groups such as NFPA and ANSI try to resolve conflicts between their standards, but at times it is fruitless. If you are specifying equipment and find that there is a disagreement between organizations, you need to look at each standard and who the intended user of the equipment outlined in the standard was. NFPA standards are often based on the needs of emergency responders (such as NFPA 704).

You should also use standards for maintenance purposes, as shown in Figure 11-10. Standards often contain or reference information on how to maintain the equipment. The standard may also provide for specific tests that can be performed to ensure that the equipment still meets the performance requirements of the standard. Some standards require testing—by using either *nondestructive tests* or *destructive tests*, such hydrostatic testing of compressed gas cylinders (SCBA bottles). Periodic testing does not typically include destructive testing. **Nondestructive tests** are intended to allow you to test the equipment without destroying the equipment. Examples of nondestructive tests include load tests for ladders and annual hose tests for fire hose. In these instances the equipment is tested and then returned to service if it passes the test. Nondestructive testing follows a procedure, but the testing procedure may be simple enough that it can be done locally without a complicated setup (hose testing is a good example). **Destructive testing** takes a piece of equipment and tests it until it fails. This testing is of value in that a representative sample of the equipment is tested—not all of the equipment. It may be the only method available to provide a reliable test for that particular piece of equipment. Destructive testing causes the loss of the equipment and is not often used outside of a manufacturer's test facility or a third-party laboratory. Destructive testing typically takes place in a laboratory to ensure the ability to replicate the results from testing similar items.

> **NOTE**
> The inspection program that is in place should document the history of your harnesses and equipment and reflect the manufacturer's recommendations.

HARNESSES

Harnesses can be used by both rescuers and victims. As discussed in Chapter 8, there are a variety of different harnesses available for use depending on who the intended wearer is and how these harnesses were designed. You should choose only those harnesses that are designed for use in rescue work and that meet the requirements of NFPA 1983 and ANSI A101.4. NFPA classifies harnesses in three categories: Class I, Class II, and Class III.

9.0 DETAILED INSPECTION & MAINTENANCE LOG: SERIAL NUMBER: _____ MODEL NUMBER: _____ DATE PURCHASED: _____			
INSPECTION DATE	INSPECTION ITEMS NOTED	CORRECTIVE ACTION TAKEN	MAINTENANCE PERFORMED
Approved By: _____			
Approved By: _____			
Approved By: _____			
Approved By: _____			
Approved By: _____			
Approved By: _____			
Approved By: _____			
Approved By: _____			
Approved By: _____			
Approved By: _____			

FIGURE 11-10 A sample inspection and maintenance log. You should consult the manufacturer for its specific recommendations as well as recognized standards.

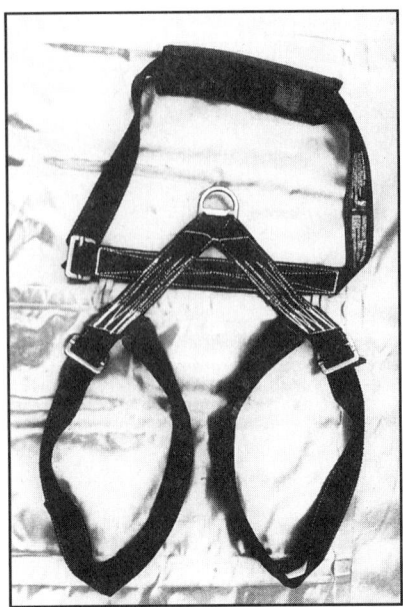

FIGURE 11-11 This is a Class I harness. It is designed to support a single person. All harnesses should be clearly marked as to their class.

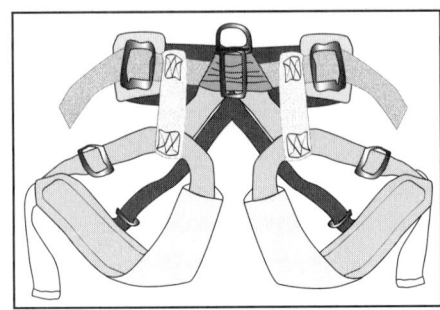

FIGURE 11-13 A Class II harness. Class II harnesses are designed to support a two-person load, and they look similar to Class I harnesses. Only by consulting the marking on the harness can you tell if it is a Class I or Class II harness.

Class I harnesses are designed to go around the waist and the thighs or under the buttocks with the intention that they will be used for emergency escape. Because Class I harnesses are designed with a design load of 300 lbf, they are intended to support only one person, as shown in Figure 11-11. Class I harnesses should not be used during confined space rescue when it is necessary or expected that a person will have to be raised or lowered using the harness. Do not confuse Class I harnesses with ladder belts. Ladder belts are intended to secure people to a stationary object where they will remain standing in an upright position, as shown in Figure 11-12.

Class II harnesses consist of a waist belt with straps around the waist and the thighs or under the buttocks that create a seat for the wearer. Class II harnesses look similar to Class I harnesses, but Class II harnesses are designed for a 600 lbf load and can carry a two-person load, as shown in Figure 11-13. Class II harnesses are worn when the person may be lowered or raised and is expected to remain upright or nearly upright at all times. Class II harnesses allow the wearer to be supported by the seat and are not designed to be used when the wearer may be inverted or rotated perpendicular to the ground. Once the wearer of a Class II harness reaches an angle that is parallel to the ground, he may slip out of it and fall. Class II harnesses can be differentiated from Class I harnesses by the manufacturer's label that is required to be permanently attached to the harness. Class II harnesses have limited use for confined space rescue.

Class III harnesses consist of a waist belt with straps around the waist and the thighs or under the buttocks and over shoulders. Class III harnesses can be constructed of one or more pieces, have a design load of 600 lbf, and are expected to carry a two-person load, as shown in Figure 11-14. Class III harnesses provide protection

FIGURE 11-12 The ladder belt worn by this firefighter should not be used to support a person while raising or lowering on a rope, cable, or other support.

Rescue Equipment 173

FIGURE 11-14 A Class III harness. This harness not only supports a two-person load, it also protects a wearer from falling out if inverted.

from falling out of the harness if the wearer becomes inverted. Depending on the location of the D-ring connectors on the harness, the wearer can also be lifted by an attachment at the shoulders, center of the upper back or midchest, or the sides and center of the waist. This type of harness is highly adaptable for use during confined space rescues. You can secure a wearer who might invert, and it is also possible to lift the wearer at the shoulders, chest, or back by attaching a line to the D-rings. The advantage to this is that when you are faced with an extremely narrow opening or an opening that requires the wearer to be hands free for using tools or equipment, the wearer can now be raised or lowered through the opening without using their hands for repositioning. If you had an 18-inch opening and the rescue entrant needed to wear an airline respirator and escape bottle, they could enter this opening while using at least one hand to align and pass the escape bottle and airline through the opening. Additionally, for an 18-inch opening, most people would have to raise their hands over their head to narrow their profile and grasp the line to pass through the opening. Trying either of these maneuvers while wearing a Class II harness could allow the wearer to fall backward if they let go of the rope to use their hands. A Class III harness keeps the body in a straight line and allows you to have both hands free.

Beyond knowing the type of harness that you are using, you must know the other limitations of the harnesses and associated equipment attached to them. What type of material is the harness constructed of? NFPA 1983 *Standard on Life Safety Rope and Equipment for Emergency Services, 2006 Edition* has specific requirements for harness design and construction. Is there a lifespan or shelf life for your harnesses? As materials age and are subjected to abrasion, sunlight, and water, they behave differently and can deteriorate after time. Harnesses that have been contaminated with oils, gasoline, grease, or other chemicals will deteriorate more quickly. You must have a documented inspection program for your harnesses (hopefully one recommended by the manufacturer) and track each harness through an identification system. You may not be able to test harnesses that are suspected of having been damaged or that may no longer be serviceable. Except for a visual inspection, it is difficult, if not impossible, to tell if a harness is damaged without destroying it. Load testing a harness usually winds up loading the harness until it fails.

> **NOTE**
> You must be aware of the limitations of harnesses and the equipment associated with them.

Other features of harness construction you must be aware of include the stitching of the harness (number of stitches per inch and the pattern), as shown in Figure 11-15. Every time a needle is passed through the fabric of a harness, it will break a certain number of fibers within the material. The more stitches per inch, the more broken fibers that can occur in a harness and thus weaken the belt. Stitch patterns are also

FIGURE 11-15 The stitching pattern and number of stitches per inch in a harness are important considerations because they can affect the strength of the harness.

important because they help resist the forces of the load being applied to the harness and spread the load over a larger area of the belt. The thread used in the construction of the harness must also be visible to the naked eye for visual inspection.

> **NOTE**
> You must initiate a documented inspection program for your harnesses and keep track of them through an identification system

Wristlets

Other items used under very specific circumstances are **wristlets**, shown in Figure 11-16. Though these devices are not harnesses, they should be treated in the same manner as harnesses, that is, you should use them within their design limits and inspect them on a regular basis. Wristlets may also be useful as a drag device when attached to the ankles during horizontal entry when the victim or rescuer cannot use a harness. Though harnesses have design loads associated with them, wristlets are more difficult to categorize. Wristlets carry the load through joints within the body and do not support the body. Because of this loading of the joints, the safety factor for wristlets is 3:1. A 3 to 1 factor means that if you intend to lift a 300-pound load, the wristlets must have a minimum breaking

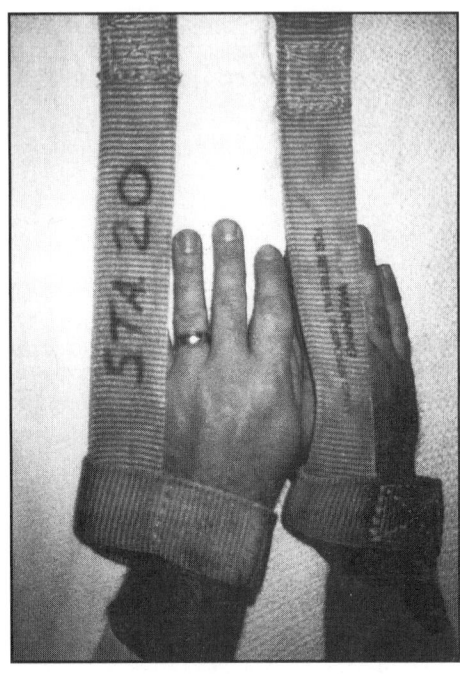

FIGURE 11-16 Wristlets that can be used to raise or lower a person.

strength of 900 pounds. The breaking strength of wristlets varies, and you must be aware of it. Few, if any, wristlets will have a breaking strength greater than 5,000 pounds.

You must consider how wristlets are intended to be used. Attaching the retrieval equipment to a person's wrists and then lifting by raising his hands over his head places all of the strain on the shoulder joints and muscles. This can be painful and may cause injury. If you intend to use wristlets (and you should consider them as part of your equipment), look at the wristlets and how they attach, how comfortable they are to wear around the wrist, and if they can be adjusted to compensate for body size. If you intend to attach them to ankles, are they large enough to go around the person's ankles? If your rescue entry involves a very narrow pipe, the rescuer must crawl in horizontally but may not be able to crawl back out. Wristlets attached to the ankles and a line attached to wristlets will allow you to drag the rescuer back out. Wristlets are a tool to use when there is no other way to get the victim or rescuer out of a confined space and haul them out. Be aware that there is the potential for injury, be aware of the limits of the equipment, and make sure that you include wristlets in your maintenance and inspection program.

Inspection

Your inspection program must document the history of your harnesses and equipment, and should be based on the manufacturer's recommendations for inspection, testing, and maintenance. The program must include all items that the manufacturer identifies as important to maintaining the reliability of the harness, as shown in Figure 11-17. Frayed material, damaged stitching, chemical exposure, and fading of the fabric due to sunlight exposure should all be addressed in an inspection program. Not only must you periodically inspect the harnesses, you must inspect them before and after each use. The pre-use inspection should involve a brief look to make sure the equipment has not been damaged in storage and should be part of your SOPs for emergency response. Harnesses contain more than fabric and stitching. There are D-rings, O-rings, snap hooks, buckles, and other connecting hardware, as shown in Figure 11-18. This hardware has specific requirements and, if it is load bearing, it must be constructed of forged, machined, stamped, extruded, or cast metal. The test for this auxiliary equipment requires that it be tested as part of the harness when it is designed to be the load bearing attachment point.

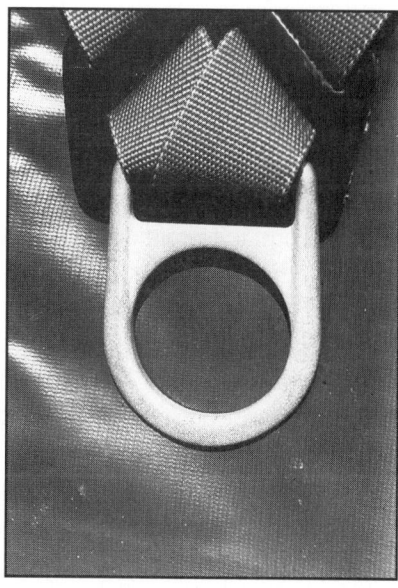

FIGURE 11-18 This D-ring is built into the harness and must be inspected as a part of the harness. The strength of the D-ring must be taken into consideration when using a harness.

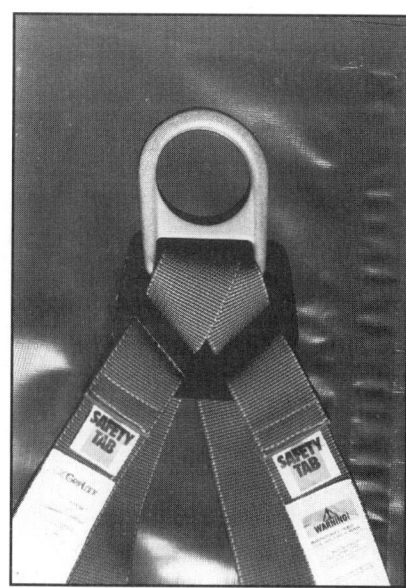

FIGURE 11-17 The safety tabs shown on this harness are an integral part of the harness and must be inspected. The manufacturer of this harness puts these tabs in to show if the harness has been impact-loaded.

As with other parts of the harness, O-rings, D-rings, and other connecting hardware must also be inspected on a regular periodic basis, before each use, and after each use. Every piece of equipment that will be used to support human life must be included in your inspection, testing, and maintenance program. Before-use inspection is the responsibility of everyone at the emergency, but to guarantee that it is inspected and inspected consistently, before anyone is put on line, you should assign the inspection as a specific task for the safety officer or other qualified person.

TRIPODS AND OTHER LEGGED RESCUE EQUIPMENT

Tripods may the one piece of equipment you picture when you think of confined space rescue, but rescue equipment is not limited to tripods. What we are really talking about is stationary equipment that will provide an anchor point to lift in a vertical position. To keep it simple in this section of the book, the word *tripod* will be used to describe all tripods and similar stationary equipment. Tripods provide a fixed anchor point for the lifting equipment. The tripod is not necessarily anchored to the ground. The tripod carries the load to the ground and when lifting with a tripod, you must watch how the equipment is loaded. All loads on the tripod, including change of direction pulleys, must be applied as axial loads to keep from tipping or collapsing the equipment. In addition to the traditional tripod, manufacturers are now making specialized tripods that include equipment with four legs and a davit arm (shown in Figure 11-19), davit arms with U-shaped bases, tripods designed to be bolted to the flange of a manway opening for a vertical confined space, attachment devices to support hoisting equipment from an overhead steel beam (as shown in Figure 11-19), and other specialized equipment for anchoring an overhead lift.

Whether it is a basic tripod or some specialized tripod, the available equipment can vary greatly in how it can be used. There will be differences in lifting capacity, use on sloped surfaces or level surfaces, the number of lines that can be attached at one time, and height of lift just to name a few. The tripod you select must be made to meet your needs and should match the way you expect to use it. This is another area where preplanning confined spaces is of value. You must match the equipment to your needs, but you should also be aware of any conditions you identify where the equipment will be inadequate. It would be nice to buy enough equipment to meet all of your needs, but that is generally not realistic. If you have identified a unique situation in which your equipment will not meet the requirements, either change the conditions pre-entry, determine if your present equipment can be safely adapted to meet the conditions, come up with another realistic plan, or decline your role as the rescue team.

> **NOTE**
> You need to look at advantages, disadvantages, and your own needs to determine if a particular piece of equipment is right for your purpose.

Lifting Capacity

Lifting capacity of hoisting equipment can vary from manufacturer to manufacturer and even between models made by the same manufacturer. At least one tripod on the market has an allowed load on the equipment that decreases as the length of rope in use increases. You must know the limitations of your equipment. Different hoisting equipment (that is, wenches) also has different lifting capacity, and you must know it. Lifting capacity can vary between models and manufacturers. Additionally, when equipment is designed to be used either on or off a tripod, make sure that you know how that hoisting equipment is meant to be used and how and where it is meant to be attached. Using the wrong attachment point or improperly attaching the equipment can lead to a connection failure and drop the people who are connected to that device.

> **NOTE**
> Know the lifting capacity of your equipment.

Surfaces

As mentioned earlier, you must maintain an axial load on the tripod at all times. Surfaces such as a sloped tank roof with a smooth surface surrounding the opening, or ground surfaces such as loose or soft soil may make it difficult to maintain an axial load. Tripods and similar

FIGURE 11-19 A sling is being used to attach a change-of-direction pulley to the rung of an aerial ladder. This type of sling and other more specialized equipment is available to attach rescue equipment to beams, ladders, and other support points.

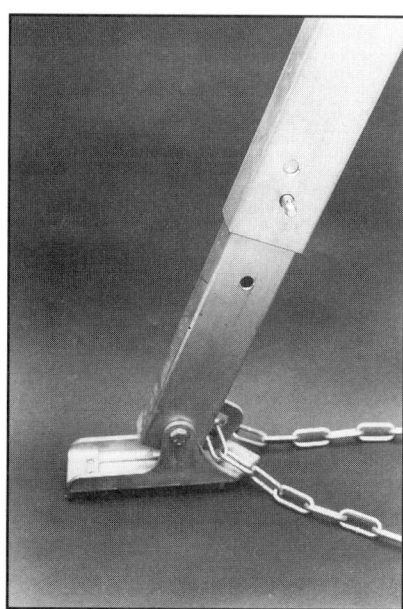

FIGURE 11-20 The foot of this tripod is designed to swivel so that it can be used flat on hard surfaces (as shown) or flipped up so that the pointed end can be pushed into soil and other soft materials.

FIGURE 11-21 Tripods that are to be placed on a sloped surface should have two of the legs placed on the same plane downhill of the opening to provide the greatest stability.

equipment may have feet that are designed to provide traction on smooth surfaces and then swivel to provide pointed feet on the reverse end that allow the foot to be pushed into soft ground for anchoring, as shown in Figure 11-20. Your preplanning activities should help you to identify some of these special needs. If warranted, you may decide to acquire special equipment for a potential rescue at that particular confined space. Before you purchase specialized equipment, determine if it might be better to change the confined space conditions (before work is started) so that your current equipment will work effectively. If you cannot make pre-entry changes to the space, safely adapt your equipment, or come up with another viable plan to make rescue entry, you may want to decline your role as the rescue team. Remember the difference between possibilities and probabilities. What is possible may not be likely to occur. Prepare for what is probable and plan for what is possible.

Using a tripod on a sloped surface presents hazards in addition to the footing. To load the tripod safely, on a sloped surface, two legs of a three-legged tripod should be placed on the same plane below the opening, as shown in Figure 11-21. If only one leg is placed downhill, the load will pivot around the single load and the tripod will have an eccentric load placed on it. The torque the eccentric load creates can cause the tripod to fall over. When using a tripod on a sloped surface, the hoisting equipment should be placed on the uphill leg of the tripod. This creates a stable base below the opening, and the application of the load on the uphill leg will assist in anchoring the uphill portion of the tripod. A sloped surface highlights the importance of properly loading the tripod. You must imagine an imaginary line that runs perpendicular to the head of the tripod and the earth. This line is the **axis** or position on which the load must travel. If the axis does not remain perpendicular between the earth and the head of the tripod, your tripod can tip over. Proper loading of the tripod does not end at the head or base of the tripod. The load carried down each leg of the tripod must be transmitted in an axial manner. If the load is eccentric, you can rotate and tip the tripod or cause the legs to collapse. Keep the load spread equally between all three legs so that it can carry the full load it is designed to support. Figure 11-22 shows a change of direction pulley used to maintain an axial load on the tripod and legs. If you attempted to pull this load out of the confined space without the change of direction pulley, the pull would be in a

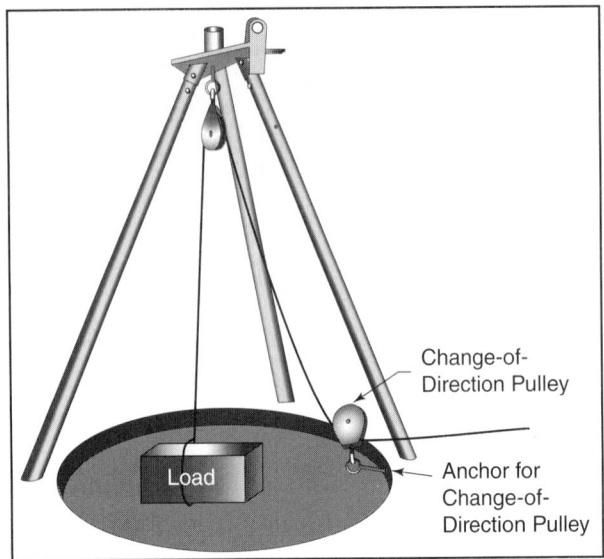

FIGURE 11-22 The load shown in this drawing is being applied axially to the tripod by using a change-of-direction pulley between the legs of the tripod. If the change of direction pulley were outside the feet of the tripod, the load would not be axial, and the tripod would tip over.

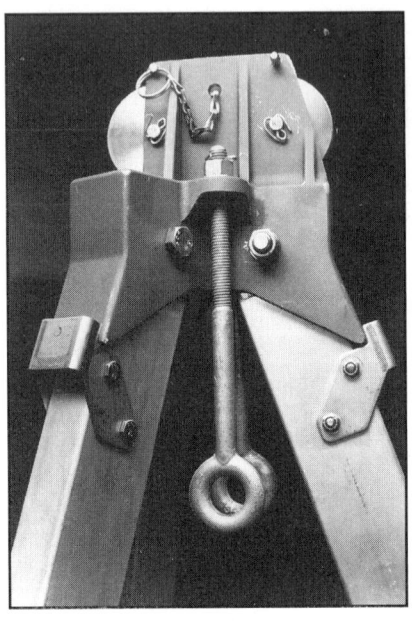

FIGURE 11-23 The locking device at the head of this tripod keeps the legs rigidly in place. If the tripod was tipped and there were no locks, the legs would pivot to the center of the tripod, and the tripod would collapse.

horizontal direction, would be applied to only one or two legs, and would pull the tripod over. With the change of direction pulley, the load remains vertical and spread between the legs of the tripod.

Tipping a tripod over can have a compounding effect. If the legs do not have a locking mechanism that locks the legs in the open position at the head of the tripod, as shown in Figure 11-23, tipping the tripod, even slightly, can allow the unloaded legs to swivel into the closed position. When the tripod is pushed back into the normal position, the other two legs are out of position and cannot take the load. The result is that the tripod falls over and the person on the tripod is dropped. Though locking the legs will not stop the tripod from tipping, it will keep the legs open and when pushed back into place, the tripod will have all three feet spread out on the ground.

Now that we've looked at the top of the tripod legs, let's look at how the legs are connected at the bottom of the tripod. The laws of physics want the legs to spread out and flatten to the ground. A method of anchoring the legs together such as a chain, shown in Figure 11-24, between all of the legs of the tripod stops this from

FIGURE 11-24 Chains between the tripod feet keep the legs from spreading as a load is applied.

happening. Without some method of anchoring the legs of the tripod, any forces spreading the legs apart are transferred to the head of the tripod and then multiplied due to the lever effect of the load being transmitted up the legs. If your tripod came with anchors and locks for the legs, use them.

Confined space rescue equipment is still evolving, and new equipment will be designed to meet specialized needs. Among the most difficult places to put a tripod is on the sloped roof of a tank or between several narrow and closely spaced confined spaces. Of course these spaces should be preplanned whenever possible, but the results of your preplanning may warrant that some type of permanent equipment be mounted in the area. You've read about changing the pre-entry conditions. As the designated rescue team, you have some say over the conditions at the confined space and insisting on changes that simplify or expedite rescue operations may be one of the best places where you should use that say. These improvements might include changes to the entry point or specialized equipment such as a device called a **transformer retrieval support**. The support is designed to be bolted directly to the flange of a manway opening, as shown in Figure 11-25, and provide a stable tripod for both worker entry and recue entry.

Tripods with height-adjustable legs may have different load capacities depending on the extend length of the leg. The further the tripod leg is extended, the less weight it can carry. Read the manufacturer's literature about your equipment. Know if extending the legs changes the load

FIGURE 11-26 Tripods with adjustable legs may have reduced load carrying capacities because the legs are extended. You must know how the performance of the tripod is affected by raising the legs.

capacity, and do not overload your tripod as you extend the legs, as shown in Figure 11-26. If extending the legs affects the tripod, stay within the manufacturer's limits.

How many retrieval devices can be attached to a tripod, as shown in Figure 11-27, will also

FIGURE 11-25 This transformer retrieval support is specifically designed and built to be bolted to the manway opening. It has a specific use and is a valuable device where the use of a tripod would be limited. (Photo courtesy of DBI/SALA.)

FIGURE 11-27 The number of retrieval devices that can be attached to a tripod will have an impact on your rescue operations. This tripod has two devices attached.

affect your rescue capabilities. The incident commander is responsible for protecting the safety of the emergency responders and must manage the risks. Rescuers who must enter the confined space are assuming additional risk, and that risk can be managed. All rescuers entering a confined space should be wearing retrieval lines. If a rescuer enters the space and needs to be rescued, you must be able rescue her quickly and from outside the space. Using retrieval lines for a rescue entrant expedites rescuing that person. To do this, you need one retrieval line per rescue entrant. In addition to the rescue line, consider using a safety line for all vertical rescues while a person is supported on the line and being raised or lowered. That means you need one retrieval line per person and one safety line per lift, as shown in Figure 11-28. When two rescuers enter (one at a time), you need a minimum of three lines available—one for each rescuer's retrieval and one separate safety line for lifting. This requires three attachment points for hoisting on the tripod or other anchor point. How many attachment points does your tripod or other equipment have? Too few attachment points don't mean your equipment is inadequate, it means you must manage the use of the lines.

Managing how many lines are in use at one time, as well as who is on lines, allows you to have two rescue entrants in the space and still rescue the victim, all using only three lines. Each rescuer enters, one at a time, and detaches the safety line (all rescue personnel in the space should keep the retrieval line attached). You then package the victim and prepare for the lift. Before removing the victim, one rescuer is removed from the confined space, and both lines from the rescuer who is now out of the space are sent back into the confined space. The remaining rescuer then attaches the victim to the safety line and the available retrieval line. Once the victim has been removed, the safety line is sent back in for the remaining rescuer, and that rescuer is removed from the space. This line management makes sure that no rescuer is without a line for emergency retrieval, and no one is supported by only a single line during entry or retrieval. You have to consider not only the potential failure of a rope or cable, but also a failure of the mechanical parts of one of the retrieval devices. Two lines build in redundancy and minimize the effects of a single failure.

Retrieval devices that use four legs and a davit arm are not actual tripods, but these devices must be loaded with as much care as a tripod. When using devices with four legs and a davit arm, you must keep the base as close to level as possible. The devices transfer the load down the davit arm, to the base, and then equally through each leg to the ground. If the device is tilted or the load is not applied axially, it may tip over or be pulled away from where it has been set up, as shown in Figure 11-29. As with any other retrieval device, you must know what limitations the manufacturer has placed on the use of its equipment. Placement on a slopped surface will require that the equipment be placed and loaded properly, and you should get specific instructions from the manufacturer.

How high of a lift you can make with a tripod, as shown in Figure 11-30, will depend on the clearance between the bottom and top of the tripod. After a victim or rescuer has been brought to the opening of the confined space, will there be enough room to completely lift him out of the opening by using the retrieval equipment? If the rescue stretcher is 6 feet long and the lifting height from the opening is 6 feet or less, will you have enough room to effectively get that person completely out of the opening by using the retrieval equipment? Or will you have to lift them as far as possible and then wrestle with the

FIGURE 11-28 The use of a safety line by rescuers should be mandatory. A safety line allows the rescuer to be protected in the event of a failure of the main line and allows the rescuer to remain attached to a retrieval line while working in the confined space.

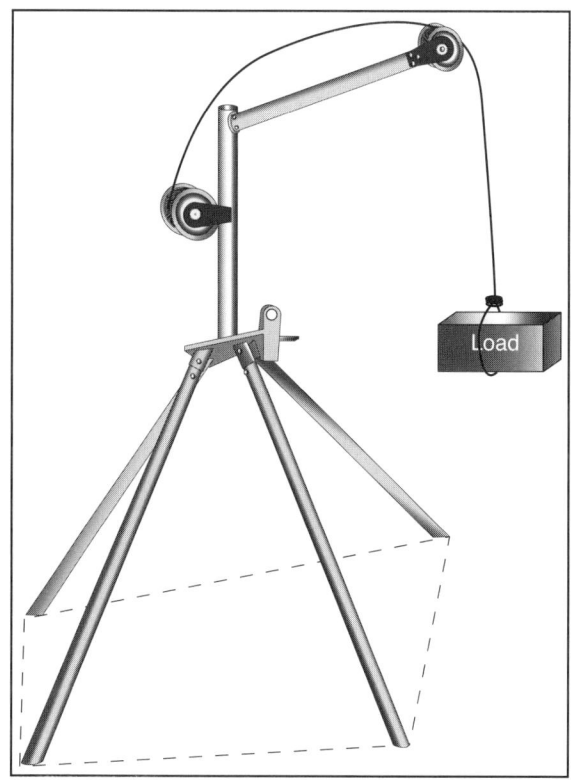

FIGURE 11-29 Just as a tripod must be loaded axially, so must other devices. This illustration of a davit arm device shows how the load must be kept within the four legs of the device to prevent tipping or other failures.

FIGURE 11-30 The height of the lift that a tripod or similar device can provide should be among your considerations when selecting the device.

stretcher to get it clear of the opening? Though this may not seem like a critical factor while you are reading this book, how do you handle a 300-pound victim without the mechanical advantage provided by your hoisting equipment or wrestle a rescue stretcher on the top of a narrow, sloped roof tank? Know how much clearance you have with your tripod once you get the victim out of the space. If it is inadequate, determine how you can overcome that problem and work your plan.

HOISTING DEVICES AND FALL PROTECTION

There are many different ways to raise victims or rescuers out of a confined space. Your equipment should be efficient to use, provide mechanical advantage and safety, be simple to operate and allow you to maintaining proficiency with the equipment. Some of the simplest equipment may be mechanical winches designed for confined space rescue, but the simplest may not always match the circumstances. Depending on your knowledge and proficiency, rope, pulleys, carabiners, and other types of rope equipment may be best to use. This book is not about using rope and other rope rescue equipment such as pulleys, carabiners, etc. for confined space rescue—that information is beyond the scope of this book. If you decide that the use of rope rescue equipment, techniques, and procedures for confined space rescue (commonly called high-angle rescue) is best for your use, you should take an additional course of study in rope rescue. Rope rescue demands a high level of initial training, and you must be able to maintain proficiency through continuous training.

Retrieval winches are designed for hoisting people and may have stainless steel or galvanized steel cables attached to allow them to be used as lifting devices. The winch provides a mechanical advantage for raising and lowering people. (Please note that not all retrieval winches can be used for controlled lowering—some can be used only for raising.) In addition to the mechanical advantage, other built-in features include fall protection, handle brakes to prevent movement of the load when you release the handle, the ability to adjust the mechanical

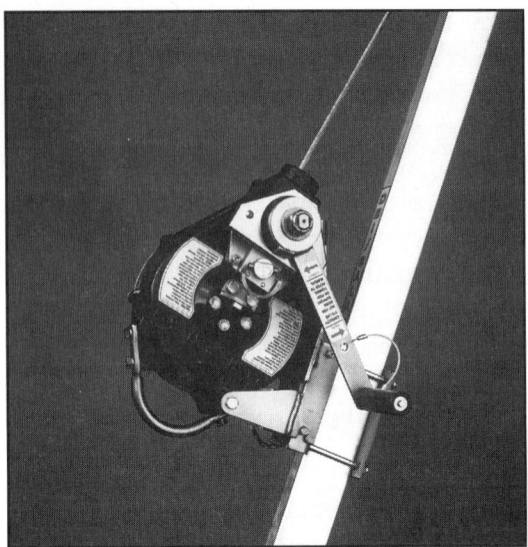

FIGURE 11-31 This retrieval device has a variety of features that make it valuable for confined space rescue. In addition to being useful in several different positions, it can be used as an untended safety line because it provides fall protection.

FIGURE 11-32 A snap hook for connecting O-rings, D-rings, and other equipment to retrieval equipment.

advantage provided, and a clutch mechanism (shown in Figure 11-31) to prevent applying the force of the mechanical advantage on an entangled person. Limitations to the use of retrieval winches include the length of the cables, periodic inspection requirements that mandate that some units be returned to the manufacturer for routine inspection and maintenance, the cost per unit, the need to buy a separate unit for training use, and manufacturer's requirements that the a unit be sent back to the manufacturer for inspection after each rescue use. As with any other equipment you obtain, you must look at the advantages and disadvantages of the equipment and match it to your own needs.

A winch with a built-in self-retracting safety line, shown in Figure 11-31, can be a great advantage during confined space rescue. If you were faced with a limited number of team members and needed to maintain both a safety line and a retrieval line, it could be difficult to maintain the proper amount of slack on the safety line while raising or lowering the person with the retrieval line. Using a self-retracting safety line allows the line from the winch to freely extend or retract as fast as the wearer moves. If the wearer moves too quickly (such as accelerating during a fall) the line locks and stops the fall. Self-retracting winches limit a free fall to 18 inches or less.

A self-retracting line used during raising or lowering operations is one less line that needs to be tended by a person.

Snap hooks, shown in Figure 11-32, allow you to connect D-rings and O-rings to the life line. Snap hooks are available with either spring-loaded or double-locking closures, as shown in Figure 11-33. You should use double-locking snap hooks only for rescue operations and make sure you match the size of the snap hook to the

FIGURE 11-33 Match the snap hook to the size of the device to which it is to be connected. The snap hook on the left is a spring-loaded snap hook, and the one on the right is a double-locking snap hook. Double-locking snap hooks are the only type of snap hook that should be used to support human life.

size of the attachment to which it will be connected. Using a snap hook with a throat opening that is too large can allow an O-ring or D-ring to either side load the gate or slip under the keeper of the gate and slide out of the snap hook. As with any other load, the load that is applied to the snap hook must be along the major axis of the snap hook. Loading the keeper is not recommended because the keeper is tested to withstand only a 350-pound side load. Applying a load greater than 350 pounds sideways to the keeper of the snap hook can deform the keeper and leave an opening where the keeper was. If the throat opening of a snap hook is greater than 1 inch, a locking feature should be included in the design. Like all the rest of your equipment, snap hooks must also be inspected and maintained. Follow the manufacturer's instructions and look for corrosion of the metal, distortion of the snap hook and keeper, and the functionality the swivels, springs, and other movable parts of the snap hook.

Protect the integrity of your rescue equipment. Using rescue equipment to raise or lower equipment, tools, or materials into or out of a confined space can damage the equipment and should not be allowed. Rescue equipment should be reserved for use only to support human life. Yes, you can use the rescue equipment to move people into or out of a confined space during routine work practices, but raising or lowering equipment or tools can overload the tripod, winch, or other apparatus associated with the rescue equipment. The damage caused by the overload may not be visible or obvious, and the equipment can fail without warning. Equipment used for rescue purposes must be capable of performing to its design limits during an emergency. The demands placed on the equipment and the need for total reliability mandate so many standards for rescue equipment. There is a certain expectation that equipment used during rescue will be subject to extraordinary loads so that it will not be easily overstressed and fail. Using hoisting equipment that has been used to lift tools and equipment and may have been overstressed can cause the failure of the equipment and injure or kill both the victims and rescuers. You may see claims by some manufacturers that their tripods and associated equipment can be used for hoisting people and equipment. How do you know whether the equipment had too large a load on it? Equipment not dedicated solely to supporting human life can be used for anything. Did someone borrow a tripod or hoist and use it to lift an engine out of a car? After all, it is an equipment lift too. Not everyone appreciates the importance of this equipment and knows how to use it properly. Though this problem is difficult to manage, consider the consequences of not managing it. If you plan on managing the risks associated with the rescue problem, you must start before the emergency and make sure that your equipment can be used with confidence.

> **NOTE**
>
> Hoisting equipment that has already been overstressed by lifting tools and equipment can fail because of the demands of an emergency and injure or kill both the victims and rescuers.

Ropes and Rope Equipment

Using rope and rope equipment for confined space rescue is fairly common. However, for some emergency responders, using rope and the associated equipment may be beyond their capabilities. Yes, you can attend and pass an initial rope rescue course, but can you maintain the proficiency that you will need during an incident? A skill as simple as tying knots correctly can be lost if you do not practice it. Combining knots with equipment and creating systems is a much more difficult skill to maintain. The use of rope for confined space rescue must be limited to those people who can acquire and maintain the proper skills to use their equipment correctly. They must know what knots to use and how to tie them, and they must be able to identify the limits of competency. For confined space rescue, you must consider if using simple basic equipment and tools that require minimal training is better than using rope. If the rescuers cannot train on a regular basis and maintain skill proficiency, you would be better off to plan to use an outside trained team. Even when you train on a regular basis, you may not be able to handle every rescue situation. Know when to call for help and who to call. This information can be as valuable as having your own team. The best advice that this book can offer on the selection and maintenance of ropes and rope equipment is to refer to **NFPA Standard 1983**—*Fire Service Life Safety Rope and System Components*. This NFPA standard is an evolving document that has been developed and revised by a technical committee of users, manufacturers, and special experts. It covers rope as well as associated equipment such as harnesses, carabiners, snap-links, and descending and ascending devices.

> **NOTE**
> Tying knots correctly is an invaluable skill, but one that can be lost if you do not practice.

The reuse of life safety rope after a rescue is still controversial. NFPA Standard 1983 (2006 edition) provides guidelines regarding the reuse of life safety rope. The manufacturer must provide guidelines on the reuse of rope, and rope should be considered for reuse only if *all* of the following conditions are met:

- Rope has not been visually damaged.
- Rope has not been exposed to heat, direct flame impingement, or abrasion.
- Rope has not been subjected to any impact loads.
- Rope has not been exposed to liquids, solids, gases, mists, or vapors of any chemical or other material that can deteriorate rope.
- Rope passes inspection when inspected by a qualified person following the manufacturer's procedures both before and after each use.

> **NOTE**
> The reuse of life safety rope after a rescue is still controversial.

Determining who is a qualified person is also controversial, and the best definition available is from ANSI. They define a **qualified person** as "one who by possession of a recognized degree, certificate, or professional standing, or by extensive knowledge, training and experience has successfully demonstrated the ability to solve or resolve problems relating to the subject matter, the work, or the project." Simply because a person holds a higher rank does not make them qualified to inspect a rope or rope equipment. Anticipate that whenever you use your life safety rope for a rescue, you will destroy it and replace it with new rope. It is easier to expect that you will have to replace the rope than to try and convince people it needs to be replaced.

This set of conditions is difficult to meet for a reason. What are the consequences if the rope fails? Confined spaces come with many different hazards—both physical and chemical. Cuts or abrasions from sharp edges around the opening of the space can damage the rope. High heat within the space or chemicals in the space can cause deterioration. Using rope for confined rescue brings its own set of problems: You must be able to maintain not just the rescuers' capabilities, but the rope and associated equipment. This is not an attempt to discourage you from using rope for confined space rescue. It is invaluable in the hands of well trained rescuers, but it increases the complexity of your rescue operations as well as your emergency services organization.

Look at NFPA 1983 for specifications regarding the original purchase of as well as inspection, maintenance, and testing of the equipment. All associated equipment such as carabiners, snap links, and O-rings, must be maintained so that they are free of corrosion, so that parts are operating smoothly (as shown in Figure 11-34), so that parts are not deformed for any reason, and so that the equipment is free of grease, oil, and other contaminants. Establish a record-keeping system for equipment inspections and maintenance. When the reliability of equipment is in doubt or equipment is unusable, take it out of service. Have it properly tested by the manufacturer or properly dispose of it. Maintenance must be done on a regular basis, and all equipment must be inspected periodically, before each use, and after each use. Equipment that has been dropped or impact loaded may be unserviceable or require inspection and/or testing before being returned to service.

Talk to the manufacturers for specific recommendations regarding inspection and testing of equipment. Discard equipment by destroying it so that others cannot use it. If intact harnesses are thrown out in the trash, someone can take those intact harnesses and use them. The people

FIGURE 11-34 Maintaining your equipment in a safe manner is essential to the reliability of the equipment. The duct tape shown here is not an acceptable repair to the damaged lock on this snap hook.

taking the used harnesses from the trash may not know they are unusable or they may not know-how to use them properly. There have been cases in which equipment was not properly disposed of, the equipment failed, and people were injured. The previous owners and the manufacturers faced lawsuits. NFPA 1983 requires that equipment removed from service be destroyed.

PEOPLE AND EQUIPMENT

Earlier chapters in this book have discussed many different types of equipment—monitoring equipment, lockout/tagout equipment, ventilation equipment, and so on. But equipment is useful only if the people who use it know how to use it correctly. Using equipment correctly means that you understand the purpose of the equipment, the design features, and how to maintain it. Most of us know how to drive a car, but how many of us could drive a high-performance race car well? If you looked at the instrument panel for a race car, would you understand the gauges and be able to interpret the information they provide? Monitoring devices are our instrument panels. We must know how to interpret the readings and when we can rely on that information. If we fail to calibrate and maintain that equipment, something as simple as a low battery could bring everything to a halt. What could be simpler than lockout/tagout equipment? When is the last time it was inventoried to make sure that all of the equipment is there and that it functions properly?

Part of any maintenance program should be cleaning the equipment. Cleaning is more than making the equipment look brand new; it familiarizes personnel with the equipment and allows an in-depth inspection during the process. You should not limit your maintenance program to just the most obvious pieces of equipment. You must include accessories such as power cords and attachments that enhance the usability of the equipment. If your ventilation equipment requires electrical power, how well do you maintain the electrical cords and connections? Damaged or frayed power cords should not be kept in service. At the very least they can fail, you lose power, and the fan stops working. At worst damaged cords and connectors can cause electrocution or be an ignition source. Damaged equipment of any type has no place at an emergency scene. It can fail and create hazards to rescuers or victims. Competent rescuers must be able to look at equipment, determine its serviceability, and make a decision that it is or is not safe to use. They must be authorized to take equipment out of service and know how to get it repaired or replaced.

A trained professional rescuer is key to the proper use of confined space rescue equipment. Not only must they be given the skills and training to do their jobs, they must maintain and build on those skills. The initial training may provide comprehensive basic classroom training and limited field training. Training and knowledge retention is enhanced when students can apply what they have learned. Hands-on training and repeated practice allow you to develop a working knowledge of the tools and equipment used during a confined space rescue. That working knowledge supports safe and efficient operations. SOPs support your training, help you to retain your knowledge, and provide the basis for your tactical operations. Well informed trained rescuers also allow the incident commander to implement an action plan without having to give exacting details regarding what needs to be done. Preparing personnel to respond to an emergency through ongoing training programs provides a skills maintenance plan.

Document your training, evaluate your operations through training, and maintain proficiency. Do not let your experience base get ahead of your knowledge base. Both your level of experience and your level of knowledge must stay in agreement. If you come across something different or new during training or an incident, record it and evaluate what you found, what caused it, and how it affects your operations. If you fail to identify and understand the problem, you are letting your experience get ahead of your knowledge. Just because you were successful the first time the problem occurred, if you do not understand the problem, it may cause you to fail the next time it happens.

> **NOTE**
> Training and evaluating skills and knowledge verifies the competency of your emergency personnel.

LESSONS LEARNED REVISITED

The worker in this incident died of multiple blunt injuries, yet he fell only 12 feet. The platform also fell but did not strike the worker, and it was on the top of the forks because the victim had lifted the platform from the side rather than the front. He was also not authorized to operate the forklift but had done so. The equipment was used improperly and personnel were not adequately trained.

Discussion Questions
1. What NFPA or ANSI standard should the harness meet?
2. Would you consider having a qualified person check equipment that would be used to support human life (such as the work platform) prior to using the equipment?
3. Who would that person be? The incident commander, the operations officer, the safety officer, or another individual?

SUMMARY

- All resources for confined space rescue have limitations, and you must know and understand those limitations.
- Know what you want to accomplish before you begin specifying equipment for purchase.
- Whenever possible, use nationally recognized standards as the criterion for your resources. This criterion can include equipment and the qualifications of personnel.
- Match your resources to your incident conditions, when necessary call for help, and do not exceed the limitations of your resources.
- Though people are not considered equipment, they are resources.
- Inspect, test, and maintain your resources to ensure the reliability of those resources.

REVIEW QUESTIONS

1. An axial load is a load that is transmitted through the axis of the object supporting the load. If a tripod is not loaded axially, how could it fail?
2. Impact loads result from the acceleration force being applied by a load in motion. Give an example of an impact load and how the speed of the application of the load can give different results.
3. What is the value of standards when specifying equipment for confined space rescue?
4. If you have an older piece of equipment that was designed to a recognized standard and an updated version of the standard is created, will your old equipment automatically meet the new standard? Explain your answer.

5. What are the three different classes of harnesses defined in the text? How do they vary in their use and applicability for confined space rescue?
6. What is the value of having legs that lock to the head of a tripod? Explain your answer.
7. How can you use three lines attached to hoisting equipment to send two rescuers into a confined space? Why would you use three lines instead of two?
8. If you were called to the scene of a confined space accident and the strategic factors of the emergency proved to be beyond your ability to operate, how could you handle the emergency? Explain how pre-incident information would help you prepare.
9. Why must equipment be inspected periodically, after each use, and before each use?

KEY TERMS

static load 166
impact loads 166
working loads 168
axial loads 168
eccentric loads 168
nondestructive test 170
destructive test 170

wristlets 174
axis 177
transformer retrieval support 179
Retrieval winch 181
qualified person 184
NFPA Standard 1983—*Fire Service Life Safety Rope and System Components* 184

ACTIVITIES

Choose one piece of rescue equipment (for example, a tripod, a meter) and review the manufacturer's instructions for that piece of equipment. Make a list of operating limits or instructions that you were not aware of or that are new to you for that individual piece of equipment. Rate the critical nature of each of those items on a scale of 0 to 10. A 0 rating means that there is no impact and a 10 rating means there is an imminent life safety hazard. Determine how you will address each item.

ADDITIONAL RESOURCES

Manufacturer's instructions for each piece of equipment, including the correct model and/or unit number.

Applicable NFPA or ANSI standards. Look at the product label for applicable standards.

12 Team Evaluation

LESSONS LEARNED: WORKER SUFFERS HEART ATTACK IN A WATER STORAGE TANK

A 60-year-old construction company worker suffered a heart attack while working on a new water storage tank at the Acme Chemical Company plant. The worker was removed from the scene by EMS with assistance from the fire department and transported to the county hospital. This is the first accident at the site for this new plant, which when it opens will be the county's largest employer.

EMS workers, who arrived before the fire department, climbed down into the tank by way of an interior ladder and began treating the worker. Removing the worker from the tank was complicated by the fact that access was a 26-inch opening on the top of the tank and was about 20 feet off the ground. The fire department, which had recently attended a high angle rescue class, was able to use ropes and a tripod to pull the worker up from the space and then lower him to the ground.

Critical Thinking Questions

1. What qualifications do the fire department and EMS workers have for confined space rescue?
2. How long did it take the fire department to respond to this emergency?
3. Are any of the ESO workers trained in confined space rescue?

LEARNING OBJECTIVES

OSHA: 1910.146
By the end of this chapter, you should be able to:

- Perform an evaluation of a potential rescue service or team to determine if it:
  - Can respond in a timely manner.
  - Is adequately trained and equipped (OSHA 1910.146 Appendix F)

INTRODUCTION

This chapter will be somewhat unique in that the intended audience is more than just rescuers or members of a rescue team. OSHA requires employers to evaluate and select a rescue team that can reach the victim in an appropriate time frame and has the training and equipment to perform the required rescue services. Due to that requirement, this chapter is also intended to provide guidance to the people who will select a rescue team and may require that rescue team to respond to their facility.

OSHA'S RESPONSE TIME EVALUATION

On first read of the rescue and emergency services section of the standard (OSHA1910.146), OSHA uses the phrase *ability to respond to a rescue summons in a timely manner*. That is very open-ended wording because it does not define a timely manner, but leaves it up to the evaluator. Would 15 minutes be a timely manner if the victim were in a permit-required confined space with or without respiratory protection? Would 30 minutes be a timely manner for a victim having a medical emergency in a non–permit required space? Would the time of day, as shown in Figure 12-1, affect the response? There is no time frame given in the standard, but you can begin to define *a timely manner* by characterizing the hazards of the confined space.

FIGURE 12-1 This is a well-equipped rescue rig staffed by trained rescuers, but that is just one part of the evaluation. How quickly they arrive at the emergency and rescue the victim is just as critical.

Characterizing the Hazards of a Confined Space

By now it should be fairly obvious that the more threatening the atmospheric hazards and/or the physical hazards are to the victims, the more critical it becomes to quickly isolate them from the hazard. A victim in a low oxygen atmosphere wearing respiratory protection has the potential to be exposed to that atmosphere if the respiratory protection fails, loses it air supply, or is dislodged from his face, as shown in Figure 12-2. The time frame to successfully rescue that person is very

FIGURE 12-2 If the workers must enter a confined space and wear PPE to protect themselves from the hazards in the space, failure of or damage to the PPE can expose them to an IDLH situation.

short. Similarly, a person engulfed by a flowable material inside a confined space can suffocate as the material presses on the chest and stops respirations. In these cases, the hazard is imminent and rescue must be immediate to have a chance at success. What is considered a timely manner when there is an imminent hazard is different than the timeliness needed to remove a conscious and alert victim with a broken leg in a well controlled, non–permit required space? So our first evaluation criteria should be: *What are the hazards of the space, and do they create an imminent hazard to the entrants?*

> **NOTE**
> The hazards in the space and the risks to entrants should be among the first considerations you use to characterize a confined space.

The next evaluation criteria that should be used for evaluation is: *How are the hazards controlled?* The best way to control the hazards is to remove them, but that is not always possible. When the hazards cannot be removed, engineering controls should be the first choice to manage the hazards. This means lockout/tagout, ventilation, removing or limiting the amount of materials within the space at any one time (that is, restricting the flow through a space that cannot be completely shut down such as a sewer), limiting the number of people or the types of activities in the space (for example, hotwork), and any other means that separate or limit the hazards within the space. Only if engineering controls will not be effective should you consider the use of personal protective equipment. Each type of control has limited effectiveness, but the use of PPE anticipates that the engineered controls will not be fully effective. The greater the reliance on PPE for protection from the hazards, the higher the anticipation that people can be exposed to the hazard. Once exposed to the hazard, the victim is vulnerable to injury and the longer they are exposed, the greater the injury.

> **NOTE**
> How the hazards are controlled should be the next consideration you use to characterize a confined space.

Time Considerations

Now that you know the risks of the confined space and the means of controlling the hazards, you should be able to qualify how much time you have to rescue the victim. Be careful how you summarize the time to rescue the victim. It takes time for the rescue team to be notified, to respond, to set up their operation to reach the victim, and to remove the victim.

When you evaluate the timeliness of the team's response, begin with notification. How do you reach the team to let them know an emergency has occurred? If you use an offsite team such as a public fire department and call them through the normal dispatch channels, it will add minutes as the call is handled and the team is dispatched. There can be a subtle time delay in the notification process, how does the attendant call for help? If the attendant can reach the dispatch center directly, little time is lost. If he has to call a supervisor who then must call for help, time will be added to the notification process.

Response time is simply how long it takes the rescue team to get from the location they were dispatched from to the site of the emergency. The response time can vary based on traffic conditions, weather, familiarity with the site, and many other factors. You also must consider rescue team availability. If the staffing of the rescue team fluctuates, if they are at another emergency, or if they go out of service at certain times of the day, you may have to rely on another rescue team from a more remote location.

Just because the rescue team is on the site does not mean that they will be able to immediately initiate the rescue. The incident commander must size up the incident, retrieval and rescue equipment must be brought to the space, and it must be set up for use. Depending on the type of equipment needed and the location of the confined space, this setup time could be significant. Even if the equipment is staged on the site, it must still be brought to the location and set up. Something as simple as bringing a tripod to the top of process equipment located 50 feet above the ground takes time and effort, as shown in Figure 12-3.

The time to reach the victims and rescue them must also be considered. It is critical that people exposed to a toxic or low oxygen atmosphere get out of that atmosphere as soon as possible. Once the rescue team has made entry, it must locate the victims, package them, and remove them from the space, as shown in Figure 12-4.

> **NOTE**
>
> Notification time, response time, setup time, and the time it takes to locate and remove the victim should be the minimum considerations you use to evaluate the ability of a rescue team to respond in a timely manner.

FIGURE 12-3 This tripod is being carried from the rescue rig to the scene. How long will it take to get the tripod set up and operational?

FIGURE 12-4 This victim is being lowered to the ground after having been rescued. The time spent locating and removing can be reduced by having the rescue team and equipment pre-staged at the confined space.

OSHA'S POTENTIAL RESCUE TEAM EVALUATION: QUALIFICATIONS

The OSHA standard for permit-required confined spaces requires that rescue team members be trained as confined space entrants. There are also specific requirements for basic first aid and CPR training as well as training in the teams' assigned duties, PPE, and access to all permit spaces for both training and pre-planning. However, the standard requires only that at least one member of the rescue team with a current certification in first aid and CPR be available to respond.

Evaluation Components

To evaluate the rescue team, OSHA recommends a two-part evaluation (see *Rescue Team or Rescue Service Evaluation Criteria*, 1910.146 Appendix F). The first component is the initial evaluation. This initial evaluation is a meeting between the user (employer) and the team (leaders or representatives) to determine the following:

- The employer's needs based on response time
- Whether the team can respond in a timely manner
- The availability of the rescue service

- If the team is adequately trained and equipped to perform the type of rescue needed at the employer's facility
- If the team is an off-site service, whether it is willing to perform rescues at the employer's facility
- If there are adequate communications between the attendant, employer, and rescue services to call for rescue assistance
- Whether the rescue service has adequate training and equipment to perform the types of rescue needed at the employer's facility

The second component is a performance evaluation of the team. In a performance evaluation, the employer watches the rescue service perform a practice rescue or an actual rescue, as shown in Figure 12-5. In preparing for a performance evaluation, there should be specific objectives the rescue team must meet. The end result of the evaluation should not be, "it looks good to me." Measurable objectives provide an objective means of determining what the team can do. Failure to meet all of the objectives should not be reason to automatically discount the team. Depending on the nature of the failures, it may be possible to provide additional training, education, or equipment to allow the team to meet the required objectives.

The performance evaluation should verify the following:

- Members of the rescue service have been trained as permit space entrants, as required by 1910.146.
- Every team member is trained in the use of and need for the required PPE, roles and responsibilities, and specialized rescue equipment.
- Every team member is trained in first aid and CPR.
- Team members can safely and efficiently perform their tasks.
- The rescue service has the ability to test the atmosphere to determine hazards.
- Rescue personnel can use entry permits, hot work permits, and MSDSs to obtain needed information.
- The rescue service can identify hazards outside the space that might endanger personnel.
- The rescue service can package and remove victims from spaces with limited size openings, limited internal space, or internal obstacles or hazards.
- The rescue service is capable of providing high angle rescue if needed.
- The rescue service has a written plan for the types of rescues expected at the facility.

> **NOTE**
> Both the initial evaluation and the performance evaluation of the rescue team should be based on measurable objectives.

Once the rescue team has been has been evaluated and found to be adequate, there are requirements for periodic team training and access to all permit spaces so that the team can pre-plan and practice rescue operations. Remember, if the selected team is an off-site team, the team must agree to provide rescue services at the employer's facility. If the off-site team declines to provide rescue services, the employer cannot rely on it for rescue. The agreement to provide rescue services, as shown in Figure 12-6, should be in writing,

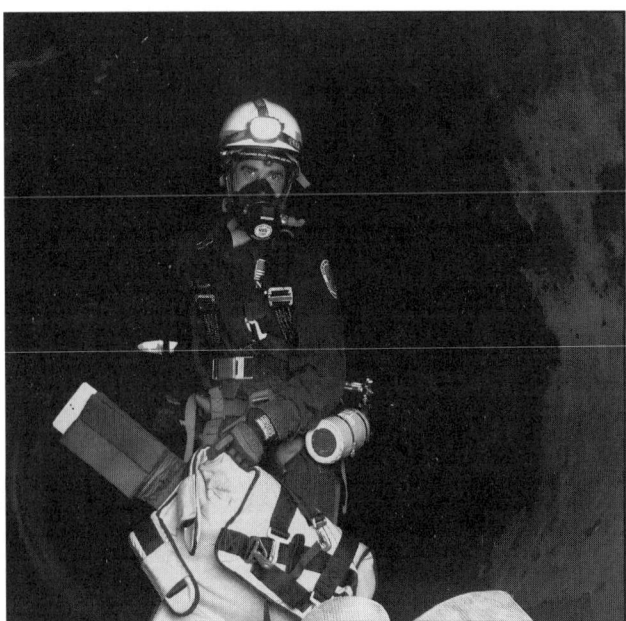

FIGURE 12-5 This is a photo of a rescuer during training. Training not only prepares you for response, it can also point out situations and conditions that might require you to revise your procedures.

XYZ CHEMICALS

Leaders in Innovative Technology

123 Main St.
Anytown, NJ 01111

XYZ Chemicals and the Anytown Fire Department mutually agree to the following in order to ensure prompt response to confined space emergencies at the XYZ Chemicals plant located at 123 Main St, Anytown, NJ

During normal operating hours (7AM to 11 PM), XYZ Chemicals will maintain an on-site confined space rescue team of at least 6 trained rescuers. This on-site rescue team will be the primary confined space rescue team for XYZ Chemicals.

The Anytown Fire Department will provide backup to the XYZ Chemicals rescue team during normal operating hours, as requested. Outside of normal operating hours the Anytown Fire Department will become the primary confined space rescue team for XYZ Chemicals. The Anytown Fire Department has received training in Confined Space Rescue that meets the requirements of NFPA 1006 and maintains a minimum 6 person Confined Space Rescue Team trained to the technician level requirements of NFPA 1670.

Outside of normal operating hours, XYZ Chemicals will not make any permit required confined space entries (planned entries or emergency entries) without first notifying the Anytown Fire Department and determining the availability of the Anytown Fire Department Rescue Team. For planned permit required confined space entries outside of normal operating hours, XYZ Chemicals shall either provide their own trained team, or they shall request the Anytown Fire Department to provide a standby team, on-site. The Anytown Fire Department will receive compensation for this off hours team based on the prevailing hourly wages of the Anytown Fire Department personnel (including overtime), plus a fee $150 per hour for each rescue vehicle required on-site. This off hours team shall consist of no less than 6 trained rescuers.

XYZ Chemicals agrees to provide access to the XYZ facility for pre-planning, familiarization and training on an ongoing basis (at least semi-annually and whenever new equipment or processes are installed). XYZ Chemicals also agrees to make its rescue team members and equipment available to the Anytown Fire Department for mutual aid within the community.

The Anytown Fire Department agrees to train with the XYZ Chemicals Confined Space Rescue Team, provide documentation of the Anytown Fire Department's training and equipment, and to submit to an evaluation of their qualifications by XYZ Chemicals.

Gregory DePaul   01/23/09   *George J. Browne*   01/23/09
Gregory DePaul   Chief George J Browne
Facilities Manager   Anytown Fire Department
XYZ Chemicals

Ed Waterson   01/23/09   *Gus S. Crist*   01/23/09
Chief Ed Waterson   Mayor
XYZ Chemicals   City of Anytown
Emergency Response Team

FIGURE 12-6 This is a sample letter of agreement between a facility and the designated rescue team. You should have a signed confined space rescue agreement that meets the requirements of your local jurisdiction.

and it should be reviewed and revised periodically. If conditions at the facility change or the capability of the team changes, the team can notify the employer and decline to provide further rescue services. Likewise, the employer can change rescue services such as establishing an on-site team if it is more effective. This agreement to provide rescue services is a two-way street. Each side has an obligation to keep the other informed when conditions change.

NOTE

An off-site rescue team must agree to provide rescue assistance.

NFPA STANDARDS

Different organizations have different audiences in mind when they write their standards. OSHA writes standards as part of the U.S. Department of Labor and has worker safety as the goal. ANSI writes standards for various audiences, including workers and emergency responders. NFPA also has specific audiences for its standards, but in many of those standards the audience is emergency responders. That is what makes the NFPA standards so valuable to rescuers—the standards are intended to address our needs.

NOTE

NFPA standards typically exceed OSHA requirements for rescue teams. As emergency responders, we should look to meet the NFPA standards.

There are at least two NFPA standards, shown in Figures 12-7 and 12-8, which you should be aware of as a confined space rescuer. NFPA 1670 is the *Standard on Operations and Training for Technical Search and Rescue Incidents*, 2004 Edition and NFPA 1006 is the *Standard for Technical Rescuer Professional Qualifications*, 2008 Edition. (All of the NFPA 1006 confined space standards have been covered in this book.) OSHA standards provide minimum requirements for

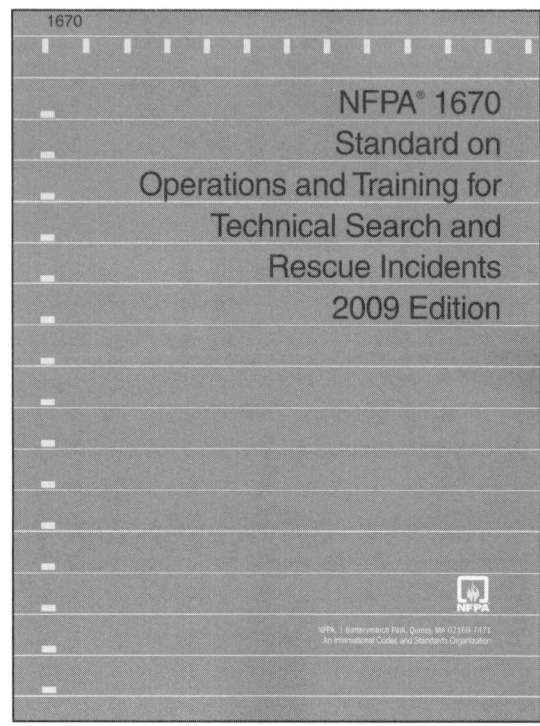

FIGURE 12-7 NFPA Standard 1670 defines the criteria for level of training and emergency operations for technical rescue teams. Confined space teams should meet the applicable portions of this standard.

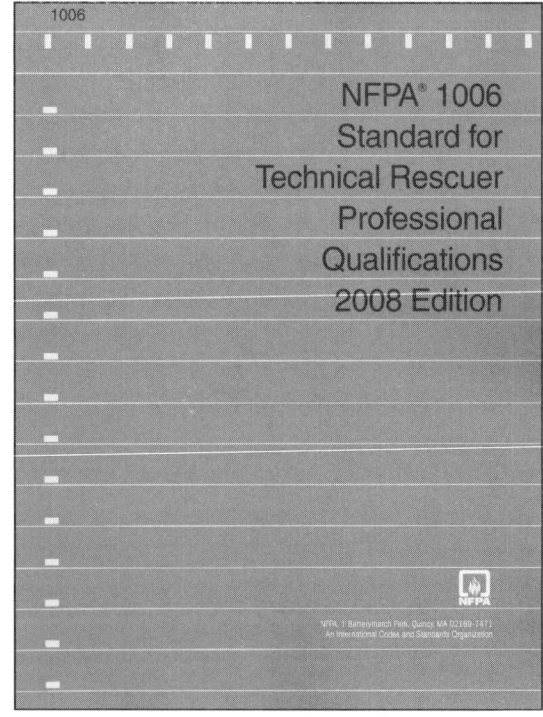

FIGURE 12-8 NFPA Standard 1006 defines the qualifications for technical rescuer. The applicable skills and knowledge in this standard should be the criteria for evaluating the individual training of members of a confined space rescue team.

compliance with OSHA regulations. You need to consider whether meeting the minimum standard is acceptable for your ESO. NFPA standards are developed by a committee of users, special experts, and other interested parties. They are updated periodically. OSHA standards often refer to NFPA standards, but they are updated infrequently. Additionally, NFPA standards contain annexes that have detailed information to explain the requirements of the standards. Look at the NFPA standards. They are developed, in part, by rescuers just like you.

NFPA 1670 addresses training by discussing several different levels of training, and it matches that training to the expected role for each level. For the awareness level, the standard limits rescuers trained to this level to non-entry functions. Under the operations level, 1670 allows entry into spaces under very specific conditions and requires that a team trained to the operations level must operate with a minimum of four people. The technician level training requirements anticipate more complex incidents and require a team working at the technician level operate with at least six members.

In addition to the requirements of NFPA 1670, Annex A of the standard provides additional information to define *timely* as it relates to confined space rescue. This annex uses many of the same criteria as the OSHA standard, but it also includes the following:

- Distance to definitive medical care. Once you have the patient out of the space, how long will it take to get them to a hospital, clinic, or other properly equipped medical facility, as shown in Figure 12-9?
- The "golden hour" principle for trauma patients.
- Recommendation of a goal of responding to confined space emergencies within 15 minutes of notification.

OTHER CONSIDERATIONS FOR EVALUATING CONFINED SPACES

Previously this chapter discussed the hazards of confined spaces in relation to the timeliness of a rescue operation. Though time is an important consideration for evaluating a rescue team and its capabilities, you need to also look at the complexity of the rescue. A complex rescue is expected to take more time to accomplish, but the more complex the rescue, the greater the demands on the skills, training, and equipment of the rescue team. Both OSHA and NFPA contain good reference for classifying confined spaces by type and the impact the classification can have on rescue. Here is that information.

In its criteria for rescue team evaluation, OSHA lists the following characteristics that may differentiate between representative spaces and worst case confined spaces:

- Internal configuration of the space
  - Open and having no obstacles, barriers, or obstructions in the space
  - Obstructed and containing obstructions a rescuer would have to maneuver around, as shown Figure 12-10
- Elevation of the opening into the space
  - Elevated with the opening 4 feet or greater above grade, as shown in Figure 12-11
  - Non-elevated with the opening less than 4 feet above grade

FIGURE 12-9 Not only do you need to consider the proximity of the nearest hospital, you must also consider if it is equipped to handle the confined space victims.

FIGURE 12-10 Rescue from this space would be difficult. Not only do you have multiple obstacles, there is very little room for rescuers to stand and work.

FIGURE 12-12 This opening is less than 18 inches in diameter and is located about 3 feet off the floor.

FIGURE 12-11 This elevated entry point to a silo creates a difficult rescue problem if it is the only access. The difficulty is increased by the snow. Would you anticipate weather conditions in identifying a worst case scenario?

- Portal size as a restriction to rescuers entering wearing PPE
  - Restricted in that the opening is 24 inches or less in the smallest dimension, as shown in Figure 12-12
  - Unrestricted in that the opening is more than 24 inches in the smallest dimension
- Access to the space and the use of retrieval lines
  - Horizontal with the opening on the side of the space; use of retrieval lines could be difficult
  - Vertical with the opening located on the top of the space, and rescuers must climb down or on the bottom of the space and rescuers must climb up, as shown in Figure 12-13

Annex H of NFPA 1670 defines types of spaces so that the information can be used for pre-planning purposes. In addition to the OSHA considerations, NFPA definitions are based on the size of the opening, the configuration, and the accessibility of the opening. The different types of openings are identified as:

- Diagonal portal
- Elevated portal
- Horizontal entry
- Manway or portal
- Rectangular/square portal
- Rounded/oval portal
- Vertical entry

FIGURE 12-13 Shown is a confined space training prop. It allows training for both vertical entry and horizontal entry. It allows an ideal location for a performance evaluation of the confined space rescue team.

Annex H also has a classification system that uses an alphanumeric code to quickly identify the space by the size of the opening, the type of opening, and rescue considerations such as accessibility. For specific information on the confined space types and classification, consult Annex H of NFPA 1670.

MANAGING CONFINED SPACES AND THE NEED FOR RESCUE

One rescue topic that has not been addressed so far in this book is the consideration for non-entry rescue. Having entrants continuously use a retrieval system would allow the attendant to rescue the entrant from the space by simply using the retrieval equipment to pull them out. The attendant is there and would not have to make entry, and the entrant would be out of the space in the time it took to use the equipment. This is an acceptable method of rescue as long as the retrieval would not increase the risk to the entrant or affect the rescue of an entrant. At the beginning of this book, the comment was made that the purpose of a confined space program is to prevent an accident. Shouldn't attendant-based rescue be considered a part of accident prevention?

Rescue Classifications

Chapter 11 contains a proposal that you consider changing confined space conditions prior to entry so that you can facilitate the use of your equipment. The intent is to modify some feature of the space (access, means of entry, etc.) so that your equipment will now match the requirements of the confined space. Installing permanent mounting platforms for a mast to hold retrieval equipment and thus standardizing the type of equipment used is one example, as shown in Figure 12-14.

> **NOTE**
> Consider changing confined space conditions prior to entry in order to reduce the hazards or facilitate rescue operations.

Let's take this idea a little further and suggest that among the goals of the rescue team and the employer is prevention as the best form of rescue.

FIGURE 12-14 A confined space shown with a permanently mounted base for a retrieval device. Simple changes such as this to a confined space not only support worker safety but can significantly improve rescue operations.

During the team evaluation, selection, and training process, both the team and the employer will be evaluating various confined spaces as well as each other. As part of this process, you should look at the confined spaces and consider ways to simplify rescue operations. If you have a permit-required confined space that will contain an IDLH atmosphere, the most significant threat to the entrants is the space. To rescue a victim from that IDLH space with any degree of success, you will want to remove them as quickly as possible. If, however, the IDLH hazard could be eliminated, you wouldn't be faced with that same sense of urgency, thus reducing the potential risk to the entrants. If the IDLH hazard can only be controlled and people will have to enter wearing PPE, you may be able to create a change whereby the entrants can be immediately rescued from outside the space by the attendant. Either way, this is a win-win situation. The entrants work with a greater margin of safety, the chances of having to enter to make a rescue are reduced, and if the team must enter to affect a rescue, their risk can be reduced.

Is it possible to take a proactive approach and either eliminate hazards or reduce the time it takes to affect a rescue? Several years back, one of the authors of this book, George Browne, worked with his colleague Marc Brodt to develop a rescue classification system managers could use to evaluate confined spaces within their facilities. This system was developed to classify the risk, identify the type of rescue, identify required rescue resources (people and equipment), and allow managers to determine if the existing conditions were acceptable. Based on the confined space classification, there were prescriptive requirements for rescue. As the hazard to the entrants increased, the more prepared the rescue team had to be for entry. In the worst case scenario, the rescue team was expected to be set up at the opening and prepared for immediate rescue entry. The purpose was to make facility managers aware of confined space requirements, develop a confined pace entry program, and develop a synonymous rescue program. It also encouraged facility managers to identify modifications that reduced the risk associated with the confined spaces and implement those changes. If the classification identified that an on-site rescue team was required, no confined space work was permitted if the on-site rescue team was not available. In the event the on-site rescue team became unavailable during the entry, the entry had to be terminated immediately. The incentive was that eliminating hazards and reducing reliance on a rescue team improved safety and made it more efficient to work in the confined spaces. Modifying the space and improving safety also reduced costs both in terms of the expense of an on-site rescue team and the use of outside contractors. Yes, there would still be some confined spaces that would require a rescue team, but by identifying them in advance, any plan for work in those spaces would also identify the rescue requirements and the rescue team.

The classification system uses four classes of confined spaces (A, B, C, and D). Class A spaces are the most hazardous and Class D spaces the least hazardous, as shown in Figure 12-15. Several steps that have to be taken to use this system are:

1. Identify all confined spaces and qualify the rescue hazard for both entrants and the rescue team.
2. Identify changes that could be made in either the confined space or the work that is to take place within the space to reduce the space classification to the lowest class.
3. Make the changes to the spaces or work practices.
4. Identify confined spaces where the hazard cannot be reduced. Determine if it is necessary to have the rescue team (on-site team or outside team) on location, set up, and in place to make an immediate rescue.
5. Label each space at the entrance with the rescue classification A, B, C, or D and incorporate it into the confined space entry program.

The definitions of each class of confined space are discussed in the following text.

Class A Confined Space: a permit-required confined space where the hazards of the space require both the immediate rescue of any victims and a rescue team to effect the rescue. An approved rescue team must be on location at the entry point. This classification is based on:

- Permit-required space
- Configuration or accessibility
- Type of hazards

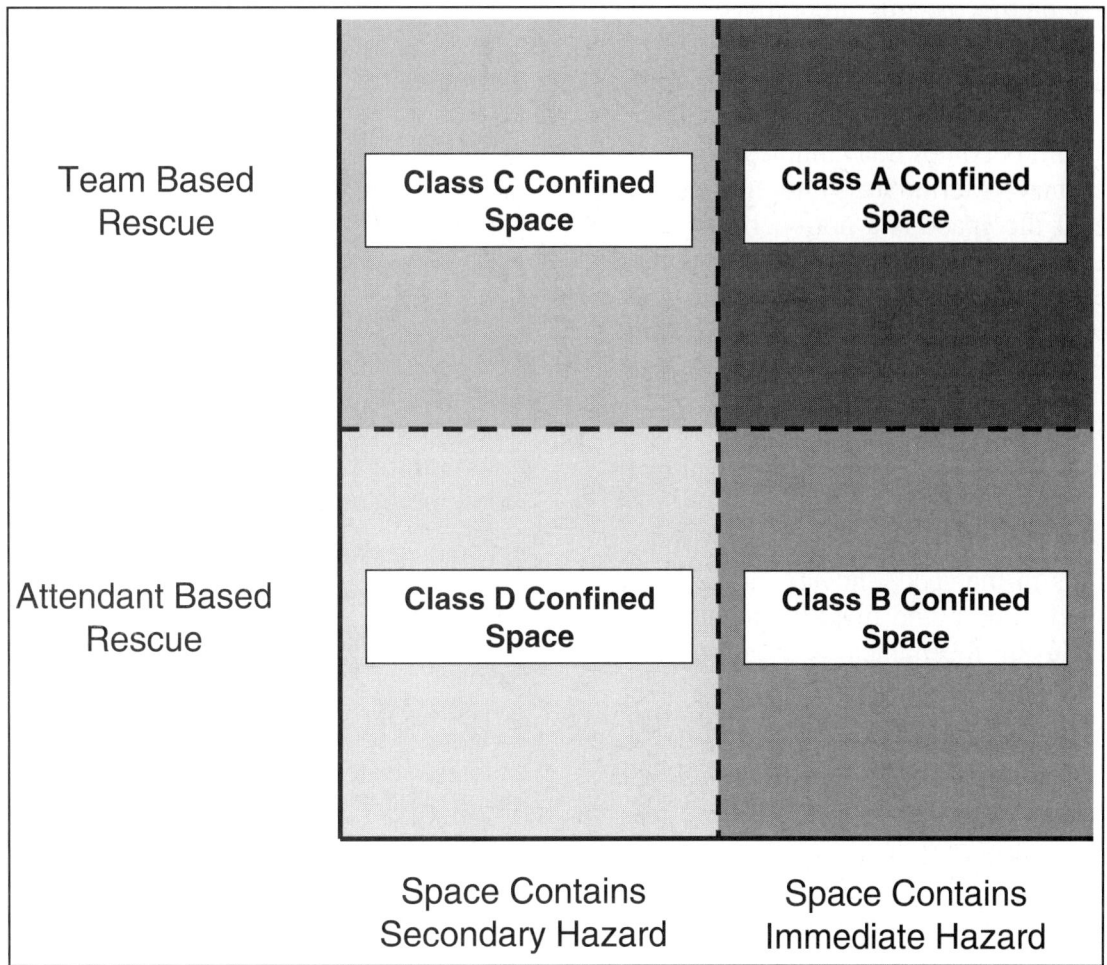

FIGURE 12-15 This is the Browne/Brodt matrix for identifying confined spaces. Immediate rescue means that conditions in the space present an imminent hazard to the victim. Secondary rescue means that conditions in the space will not adversely affect victims while they are packaged for removal from the space.

- Rescuers must enter the space to remove the victim
- The victim is in immediate peril, is unable to self-rescue, or must be physically extricated using mechanical extrication or retrieval devices

Class B Confined Space: a permit-required confined space where the hazards of the space require the immediate rescue of any victims, and the attendant can perform the rescue from outside the space. The rescue equipment the attendant uses is in place, and the entrants may not disconnect from retrieval lines. An approved rescue team must be available and able to respond in a timely manner. If the entry team becomes unavailable during work entry, the entry may be terminated and entrants may be required to leave the space. The classification is based on:

- Permit-required space
- The entrant is exposed to atmospheric or engulfment hazards
- The attendant can conduct the rescue without entering the space
- The victim is in immediate peril, is unable to self-rescue, and can be physically removed using mechanical retrieval devices

Class C Confined Space: a permit-required or non-permit required space where there are no hazards or the hazards are fully controlled, but the rescue requires the use of a rescue team. An approved rescue team must be available and able to respond in a timely manner. If the entry team becomes unavailable during work entry, the entry is terminated and entrants leave the space. If the space is a non–permit required space and the rescue team becomes unavailable, the space is immediately re-evaluated to ensure it is a non–permit space. This classification is based on:

- Configuration or accessibility
- May be a permit-required space or a non–permit required space
- The victim is not in immediate peril from conditions in the space; however, she is unable to self-rescue and must be physically removed using mechanical extrication or retrieval devices
- The entrant is exposed to the possibility of additional injury without proper packaging or treatment

Class D Confined Space: a permit-required or non–permit required space where there are no hazards or the hazards are fully controlled and the attendant can perform the rescue from outside the space. This classification is based on:

- Non–permit required space
- Entrant may require protection other than that associated with a confined space, such as fall protection, hearing protection, or other personal protective equipment
- No exposure to atmospheric or engulfment hazards that would be an immediate threat to health or life
- The attendant can conduct the rescue without entering the space

This is a simple overview of a rescue classification system for confined spaces. It is designed to protect both entrants and rescuers by identifying the rescue problem and encouraging changes to the space, prior to entry, to reduce or eliminate hazards. This is a proactive approach that aims to prevent an accident or seek ways to minimize the impact of an accident.

ORIGINALLY PROPOSED NIOSH CONFINED SPACE CRITERIA

In 1979, NIOSH created criteria for a recommended standard for working in confined spaces. The criteria were intended to address worker safety and identify the space with a classification based on the potential hazards. Four features provided the parameters for the classification:

- Characteristics of the space and rescue
- The oxygen level with the space
- The flammability of the atmosphere in the space
- The toxicity of the atmosphere

These criteria also used letters to classify the space. The NIOSH criteria are shown in Figure 12-16. Though NIOSH and OSHA did not adopt these criteria, they are valuable for comparing the differences that make a space more or less hazardous. They are included in this book for background that will allow you to better understand the disparities between confined spaces and the variance in the hazards. Today these criteria would apply to those confined spaces we now call permit-required confined spaces. The significance of this information lies in how it describes the spaces and how you might be able to update it and apply it to your confined spaces. It is of value for pre-planning by considering the oxygen, flammability, and toxicity parameters of the confined space you are dealing with and how they characterize the space. Do not use these parameters directly, but with caution. In 1979, 25 percent of the LFL was the action limit for most Combustible Gas Indicators, and entry was still considered acceptable at that level. Today that action limit is 10 percent.

Parameters	Class A	Class B	Class C
Characteristics	Immediately dangerous to life—rescue procedures require the entry of more than one individual full equipped with life support equipment—maintenance of communication requires additional standby person stationed within the confined space	Dangerous, but not immediately life threatening—rescue procedures require the entry of no more than one individual fully equipped with life support equipment—indirect visual or auditory communication with workers	Potential hazard—requires no modification of work procedures—standard rescue procedures—direct communication with workers, from outside the confined space.
Oxygen	16%* or less or greater than 25%*	16.1% to 19.4%* or 21.5% to 25%*	19.5%*–21.4%*
Flammability Characteristics	20% or greater of LFL	10%–19% LFL	10% LFL or less
Toxicity	**IDLH	Greater than contamination level, referenced in 29CFR 1910 Sub part Z***—less than **IDLH	Less than the contamination level referenced in 29 CFR Part 1910 Sub part Z***

* At sea level.
** Immediately Dangerous to Life or Health—as referenced in NIOSH Registry of Toxic and Chemical Substances, material safety data sheets (MSDS), industrial hygiene guides or other recognized authorities.
***1910.1200—Hazard Communication Standard.

FIGURE 12-16 This table is the original criteria that NIOSH used in trying to establish a classification system for confined spaces. It is used in this book only for informational purposes.

LESSONS LEARNED REVISITED

The worker was transported to a local hospital, where he was admitted and subsequently recovered from the heart attack. Both the storage tank and the chemical plant were new and unfamiliar to the rescuers. As the soon to be largest employer in the county, this plant will be an integral part of the community. The opening into the tank was 20 feet off the ground and 26 inches in diameter and should be considered a high angle rescue. The fire department has limited experience with high angle rescue, little knowledge of the types of confined spaces at this new plant, may not have been evaluated by the plant management, and may not have an agreement to provide rescue services to the plant. Would you find your self in this same set of circumstances?

Discussion Questions

1. This is a brand new facility and still under construction. Do you believe any of the local ESOs have ever seen this confined space or practiced in it?
2. How would you pre-plan large construction projects or new facilities within your jurisdiction?
3. If you began pre-planning this construction project and realized that the confined spaces were beyond your capabilities, how could you resolve that in both the short term and the long term?

SUMMARY

- OSHA provides the minimum standard for the capabilities of confined space rescue teams. Strive to meet the NFPA standards because they will better prepare your team.
- The rescue team must be available and able to respond in a timely manner.
- The team must be adequately trained and equipped to perform the types of rescues needed within their response area.
- Off-site rescue teams must be willing to perform rescues for employers.
- There must be adequate communications to call for rescue assistance.
- Be proactive and consider prevention of confined space accidents as part of the response to confined space emergencies.

REVIEW QUESTIONS

1. What is meant by this phrase: "within a time frame that is appropriate for the permit space hazard(s) identified"?
2. How is the initial rescue team evaluation valuable to the rescue team?
3. Why do practice rescue spaces need to match actual practice spaces?
4. What is the difference between a restricted and an unrestricted portal for confined space entry?
5. Identify five items that qualify a rescue team's ability to perform confined space rescues.

KEY TERMS

Class A Confined Space 198
Class B Confined Space 199
Class C Confined Space 200
Class D Confined Space 200

ACTIVITIES

Identify a confined space and its characteristics within your jurisdiction and create measurable objectives that you would use to evaluate a rescue team for this space. Consider the following:

- Response time
- Availability of the team
- Training and experience with that particular space or similar spaces
- Adequacy of the available equipment for use at the space

ADDITIONAL RESOURCES

NFPA 1670 is the *Standard on Operations and Training for Technical Search and Rescue Incidents*, 2004 Edition

NFPA 1006 *Standard for Technical Rescuer Professional Qualifications*, 2008 Edition

Glossary

Action Limit the highest percentage of combustible gas that is detected by a combustible gas detector; considered the point at which people should leave the area.

Air Purifying Respirator a form of negative pressure respiratory protection that uses filters, cartridges, or canisters to remove contaminants from the ambient air as it passes the air-purifying element.

All Risk System an Incident Command or Management System that can be used at many different types of emergencies, including but not limited to fires, police actions, emergency medical calls, and other emergencies that threaten public safety.

American National Standards Institute (ANSI) an organization that administers and coordinates a voluntary private sector standardization system. Standards developed under ANSI are consensus standards created by representatives from various interest groups.

Atmospheric Hazards the conditions which may be present in the air and that can be toxic, flammable, oxygen-deficient, oxygen-enriched, or obscure visibility.

Attendant-Based Rescue A type of rescue from a confined space; the attendant, without making entry into the space, can effectively rescue people from the space using available equipment.

Attendant the person trained and assigned to remain outside of the confined space, monitor conditions inside and outside of the space, and communicate with persons inside.

Authorized Entrants those persons trained, assigned, and equipped to enter and work within the confined space. During a confined-space rescue, those persons trained, equipped, and assigned to enter the confined space are known as the rescue entrants.

Axial Loads a load transmitted through the axis of its supporting device.

Axis the imaginary line passing through the center of a solid or plane.

Beneficial Energy Sources energy sources, the presence of which does not pose a hazard or the hazards are controlled, are of assistance during an emergency (electricity for lights and ventilation equipment, natural gas for emergency generators, etc.).

Blank Flanges flanges that have no opening in them and are meant to block the flow of a product past the flanges.

Blinding the insertion between two flanges of a device called a blind that has no opening in it and is meant to prevent the flow of a product past the blind.

Blocks energy-isolating devices meant to stop or obstruct the flow of hazardous energy or products.

Bolted Slip Blinds blinding devices that are meant to be bolted directly to a flange to stop or obstruct the flow of a product.

Braided Ropes type of rope constructed by interweaving the strands of the rope together.

Calibrated the condition of a measuring instrument after its graduations have been checked or corrected.

CALOSHA the California Division of Occupational Safety and Health.

Chains flexible series of joined links or rings, typically of some type of metal.

Chocks blocks or wedges designed to prevent motion of an object they are placed into or under.

Class A Confined Space a permit-required confined space where the hazards of the space prescribe that the required rescue is an immediate team-based rescue.

Class B Confined Space a permit-required confined space where the hazards of the space prescribe that the required rescue is an immediate attendant-based rescue.

Class C Confined Space a permit-required or a non–permit required confined space where there are no hazards or the hazards are fully controlled and the prescribed rescue is a secondary team–based rescue.

Class D Confined Space a non–permit required confined space where there are no hazards in the space and the prescribed rescue is a secondary attendant–based rescue.

Class I Harnesses harnesses designed to support a one-person load for escape purposes with the harness fastening around the waist and either under the buttocks or around the legs.

Class I, Division I, Group D electrical equipment specified under the National Electrical Code as meeting particular requirements for safe performance under certain conditions. The designation of class, division, and group refers to distinct hazardous atmospheres that may be present during the use and operation of this equipment.

Class II Harnesses harnesses designed to support a two-person load for rescue purposes with the harness fastening around the waist and either around the thighs or under the buttocks.

Class III Harnesses harnesses designed to support a two-person load for rescue purposes with the harness fastening around the waist, either around the thighs or under the buttocks, and over the shoulders to protect against inversion.

Code of Federal Regulations (CFR) the rules and regulations for different federal government departments and agencies as published in the Federal Register.

Combustible Gas Indicator a metering device intended to detect and measure the presence of a flammable gas based on how close the gas concentration is to the lower flammable limit of the calibration gas.

Command Post the location from which all incident activities are directed by the incident commander.

Communications the act of sending and receiving a message and having the message understood by the receiver.

Confined Space Supervisor the person assigned responsibility for ensuring that the requirements of a confined space program have been met prior to, during, and after entry by persons into a confined space.

Construction the materials from which the confined space is built.

Contents for confined space rescues, the materials within the confined space. These contents may be either gas, liquid, or solid and may or may not contribute to the limiting factors affecting the confined space emergency.

Corrosive a material that can be acidic or basic and, because of those properties, can damage human skin or rapidly corrode metal.

Critical Incident Stress a strong emotional or physical reaction a person may have in response to a traumatic event or incident that can affect the ability to function at work, home, or other areas of their life.

Decibel a unit for measuring sound intensity.

Defensive Operations planned actions in which there is no intentional contact with the hazards of a confined space.

Degradation the reduction of the protective properties of chemical protective clothing by mechanical, thermal, or chemical means with a loss of integrity of the garment.

Destructive Test a means of testing during which the test item is tested to failure.

Direct Reading Instruments detection and monitoring instruments that provide a reading based on a graduated scale.

Disconnect Switches an electrical switch designed to isolate the electrical source from the equipment that it powers by disconnecting the power supply from the equipment.

Eccentric Load a load applied so that the force of the load is off center of the supports carrying the load.

Emergency Services Organization (ESO) (NFPA® 1561) public, private, governmental, or military organizations that provide emergency response and other related activities. They may be for profit, not for profit, or government owned and operated.

Engulfment the surrounding and effective capture of a person by a liquid or finely divided (flowable) solid substance.

Entry Permit a written document that must be completed prior to entry into a confined space and that defines the hazards of the space, the precautions to be taken, the type of work that will be performed, the roles of personnel involved in the entry, and other specific details.

Exposures the people, property, and systems that may be affected by the confined space rescue operations.

FACE Reports fatality investigation report from the NIOSH Fatality Assessment and Control Evaluation (FACE) Program. http://www.cdc.gov/niosh/face/default.html.

Flammable or Explosive Range a definite concentration of flammable vapors in air, for a particular material, at which combustion will occur. There is a lower flammable limit and an upper flammable limit at which the vapors are either too lean to burn or too rich.

Flash Point the minimum temperature at which a material will produce enough vapors, in air, to form an ignitable mixture near the surface of the material.

Hazard and Risk Assessment determining what has happened to create the emergency, determining what conditions are still present or will evolve at the emergency, and then predicting what can be done to resolve the emergency.

Hotwork activity that creates heat, flame, sparks, or other heat sources that can ignite nearby fuels.

Hypothermia lowering of the body's core temperature.

Immediately Dangerous to Life and Health (IDLH) the maximum level to which one could be exposed and still escape without experiencing any effects that may impair escape or cause irreversible health effects.

Impact Load a load applied in a short duration so as to include the effects of acceleration in the load.

Incident Action Plan a plan developed by the incident commander that establishes goals and objectives for the emergency, identifies the resources to be used, and provides the means to accomplish the goals and objectives.

Incident Command System (ICS) see incident management system.

Incident Management System (IMS) a recognized system for providing management of personnel, resources, and activities during emergency operations by defining roles and responsibilities and standard operating procedures.

Incident Priorities the order of precedence given to the most basic goals of life safety, incident stabilization, and property conservation during an emergency operation.

Inerting the introduction of an inert gas or a gas that will not support combustion into a tank or vessel so as to exclude oxygen from the tank or vessel.

Kernmantle Rope a type of rope made with a load-bearing core (kern) and an outer sheath or braided cover (mantle).

Laid Rope a type of rope that is made by twisting smaller strands of rope together.

Latches fastening devices that consist of a bar that falls into a notch to prevent opening or operation of the object they secure.

Lead Time the amount of time that is required for a specific action to occur after notification is made of the need for that particular action.

Level A PPE the highest level of skin and respiratory protective equipment for exposure to hazardous materials.

Level B PPE the second highest level of skin and respiratory protective equipment for exposure to hazardous materials.

Level C PPE the third level of skin and respiratory protective equipment for exposure to hazardous materials.

Level D PPE the lowest level of skin and respiratory protective equipment for exposure to hazardous materials.

Liaison within the incident command system, the command staff position responsible for establishing and maintaining interaction with other agencies required to handle the incident.

Life Hazard as a strategic factor during emergency response, the threat posed to the victims, emergency responders, and spectators.

Line Valves valves in the piping that allow product to travel into a tank, vessel, vat, or other confined space.

Location and Accessibility as a strategic factor, the physical location of the confined space and the available means of gaining entry to it.

Lockout/Tagout elimination and control of hazardous sources of energy or products.

Logarithmic Scales scales, such as the pH scale, in which a change between whole numbers represents an exponent of the power to which the change will be raised.

Manually Operated Electrical Circuit Breakers circuit breakers within an electrical circuit that are designed and intended to be safely operated by direct manual manipulation.

Mechanical Ventilation the systematic movement of air through the use of fans, blowers, or other powered equipment intended to move air.

Medical Monitoring basic medical evaluation of emergency response personnel by determining and recording basic vital signs such as blood pressure, respirations per minute, and pulse.

National Fire Protection Association (NFPA) an international nonprofit organization advocating scientifically based consensus codes and standards, research, and education for fire and related safety issues.

National Institute for Occupational Safety and Health (NIOSH) the agency within the U.S. Department of Health and Human Services that identifies work-related diseases and injuries and the potential hazards of new work-related technologies and practices.

NFPA Standard 1983—Fire Service Life Safety Rope and System Components an NFPA standard developed for use in the design, testing, use, and maintenance of rope and its associated components used for rescue.

Nondestructive Test a test in which an item is tested within specific parameters to determine if it can meet the requirements of the test and recover within acceptable limits.

Non-Permit Confined Spaces confined spaces that do not contain or, with respect to atmospheric hazards, have the potential to contain any hazard capable of causing death or serious physical harm.

Occupational Safety and Health Administration (OSHA) the federal agency within the U.S. Department of Labor that is responsible for creating and enforcing workplace safety and health regulations.

Offensive Operations planned actions in which there is expected to be intentional contact with the hazards of a confined space.

OSHA, Standard 1910.146 Permit-Required Confined Spaces the specific section of the code of Federal Regulations that regulates confined spaces and the manner in which activities can occur within those spaces.

Parts Per Million (ppm) the number of units of a particular material occurring in a total volume of 1 million units. One part per million is the equivalent of 1/10,000 of 1 percent.

Penetration the reduction of protective properties of chemical protective clothing which can occur due to zippers, seams, and other openings in the protective clothing.

Permeation the reduction of protective properties of chemical protective clothing caused by the movement of the contaminant on a molecular level.

Permissible Exposure Limit (PEL) a time weighted average (TWA) concentration that must not be exceeded during any 8-hour work shift of a 40-hour work week. A short-term exposure limit (STEL) is measured over a 15-minute period.

Permit-Required Confined Spaces confined spaces that meet the definition of a confined space and have one or more of these characteristics: (1) contain or have the potential to contain a hazardous atmosphere, (2) contain a material that has the potential for engulfing an entrant, (3) have an internal configuration that might cause an entrant to be trapped or asphyxiated by inwardly converging walls or by a floor that slopes downward and tapers to a smaller cross section, and (4) contain any other recognized serious safety or health hazards.

pH Paper a basic detection device consisting of pH-sensitive paper that changes colors when exposed to an acid or base. Some pH papers are designed to change color in proportion to the pH level of the material to which they are exposed.

pH Pen a monitoring device, resembling a pen, that gives a direct reading of the pH of the material to which it is exposed.

Physical Hazards the hazards within a confined space that are produced by mechanical, electrical, chemical, or thermal means and endanger personnel in the confined space.

Positive-Pressure Supplied Air Respirator (SAR) a form of respiratory protection in which the self-contained air supply is remote from the wearer, the air is supplied to the wearer by means of an air hose, and the pressure within the facepiece is greater than the surrounding atmospheric pressure.

Positive-Pressure Ventilation a type of mechanical ventilation that creates a higher pressure within a confined space by blowing air into the space through an inlet opening with the intention of pushing the air or contaminants out of the space through an exhaust opening.

Negative-Pressure Ventilation a type of mechanical ventilation that creates a lower pressure within a confined space by drawing air or contaminants out of the space through an exhaust opening with the intention of pulling air through the space from an inlet opening.

Public Information Officer within the incident command system, the command staff position responsible for providing the press and media with information about the incident as authorized by the incident commander.

Qualified Person as defined by ANSI, a person who by reason of training, education, and experience is knowledgeable in the operation to be performed and is competent to judge the hazards involved and specify controls and/or safety measures.

Radio Frequency (RF) Interference electromagnetic interference caused by a signal generated by an electrical device.

Rescue Attendant a member of the rescue team who, through training and experience, has been qualified to be stationed outside a confined space during rescue operations to monitor authorized entrants and perform assigned duties.

Rescue Entrants members of the rescue team who have been trained to make entry into specific types of confined spaces to rescue other people who may be trapped within the space.

Resource Management the allocation and maintenance of resources required during an emergency.

Resources the people, supplies, and equipment required during an emergency.

Retrieval Winch a mechanical device designed to allow the operator to wind up or wind out a rope or cable to support a person being raised or lowered into or out of a confined space.

Saddle Vent™ a brand name for a ventilation device meant to be placed in the manway opening to allow air to be directed through the opening with minimal obstruction to the manway for the movement of people.

Safety Officer within the incident command system, the command staff position responsible for incident safety, including identifying potentially hazardous situations and the enforcement of safety procedures and safe practices.

Self-Contained Breathing Apparatus (SCBA) a form of respiratory protection in which the self-contained air supply and related equipment are attached to the wearer, and the pressure within the facepiece is greater than the surrounding atmospheric pressure.

Single Command a form of command within the incident command system when a single individual is responsible for the tasks assigned to the incident commander.

Special Problems as a strategic factor, this is a broad category of problems that are unique to a particular incident and would be likely to recur only on an infrequent basis.

Staging of Resources a method of managing resources in which those not actively being used are kept or staged in a particular area in preparation for use.

Standard 1670 *Standard on Operations and Training for Technical Rescue*—an NFPA standard intended to define and categorize technical rescue qualifications.

Standard Operating Procedures (SOPs) written guidelines for handling a specific situation.

Static Load a load applied when the load is at rest.

Time as a strategic factor, the time of day, the day, week, or year and the relative impact that it will have on emergency operations.

Transformer Retrieval Support a device that is designed to be bolted to the flange around a manway opening for vertical entry.

Tripod three-legged retrieval support device.

U.S. Environmental Protection Agency (EPA) the federal agency responsible for developing and enforcing environmental regulations.

Underwriters Labs (UL) an independent, not-for-profit, nongovernmental organization that tests and evaluates products to certify that they meet recognized standards. Products with a UL label have met the criteria for certification.

Unified Command form of command within the incident command system when more than one individual is responsible for the tasks assigned to the incident commander.

Ventilation the systematic removal and replacement of air and gases within a space.

Weather as a strategic factor, the effects that can be expected due to temperature, wind, precipitation, and other climatic factors.

Working Loads the maximum weight that a rope is expected to support.

Wristlets straps designed to be placed around the wrists of a person to allow him to be raised or lowered through a vertical opening while hanging from the straps.

Index

Note: Page numbers followed by "f" refer to Figures

A

Accidents, 2–3
Acids and bases, 39
Action limits, 34
Action plans. *See* IAP (incident action plans)
Aerial devices, 143, 176f
Agreements, 192–194
Air monitoring. *See also* Atmospheric hazards
　combustible gases, 32–36
　confined spaces, 42f
　gas-specific, 37–39, 41
　life safety and, 91–92
　oxygen monitoring equipment, 36–37
　pH devices, 39–40
　understanding readings, 40–43
Air purifying respirators (APR), 122
All risk systems, 62–63
American National Standards Institute (ANSI). *See* ANSI
ANSI (American National Standards Institute)
　equipment standards, 119, 169–170
　Safety Requirements for Confined Spaces (Z117.1-2003), 15, 19–20
　standby assistance, 17
Apparent temperatures, 110–111
Atmospheric hazards, 4–8, 21–22, 81–83. *See also* Air monitoring
Attendant-based rescues, 17
Attendants, 16–18
Authorized entrants, 18
Axes, 136
Axial loads, 168
Axis of tripod, 177

B

Bases and acids, 39
Beneficial energy sources, 50
Blank flanges, 51–52
Blinding, 22
Blocks, 51–52
Blowers. *See* Ventilation
Bolted slip blinds, 51–52
Boots, 121
Braided ropes, 143–144
Brewing tanks, 15–16
Brodt, Marc, 198
Browne, George, 198

C

Calibration, 7, 34
California Division of Occupational Safety and Health (CALOSHA), 119, 169
Carabiners, 144
Carbon dioxide, 104
Carbon monoxide meters, 37f, 38
CGI (combustible gas indicator), 32–36
Chains, 51–52
Checklists, 158, 159f–161f
Chemical contamination, 86, 122–124
Chemical protective clothing, 122–124
Chocks, 51–52
Churning, 99
Circuit breakers, 51–52
Class A, B, C, D confined spaces, 198–200
Class I, Division I, Group D equipment, 23
Class I, II, III harnesses, 119, 170, 172
Clothing, 112–113, 116, 122–124
Cold-related injuries, 114–116
Colorimetric tubes, 38, 41–42
Combustible gas indicators (CGI), 32–36
Command
　establishment of, 129, 132
　goals and objectives, 61, 63–65, 68–69
　incident commander (IC), 60–61
　posts, 65–66
　single *vs.* unified, 64
　structure of, 68–69
Communication
　between attendants and entrants, 22
　effective, 89
　equipment, 147
　noise and, 124
　terminology in IMS operations, 63–64
Confined and Enclosed Spaces and Other Dangerous Atmospheres in Shipyard Employment (CFR 29–1915 Subpart B), 10
Confined space rescues. *See* Rescues
Construction, 86–87
Contents hazards, 87
Corrosive materials, 39–40
CPR, 150
Critical incident stress, 136

D

Davit arm devices, 180–181
Debriefing, 135–136
Decibels, 124
Decontamination of chemical protective clothing, 123–124
Defensive operations, 149
Degradation of chemical protective clothing, 123
Destructive testing, 170
Direct reading instruments, 23f, 32
Disconnect switches, 51–52
Dusts, 6–7, 101
Dynamic ropes, 144

E

Ear muffs, 124
Earplugs, 124
Eccentric loads, 168
Electrical equipment, 51–53, 146–147
Emergency services organizations (ESOs)
　access for preplanning and training, 80–81

Emergency services organizations (ESOs) (*continued*)
  incident management system and, 58, 63–65, 74
  resources of, 87–88
Energy sources, 47–48, 54
Engulfment, 8
Entry permits, 15–16, 19–26, 81
EPA (U.S. Environmental Protection Agency), 122
Equipment
  axes, 136
  carbon monoxide meters, 37f, 38
  Class I, Division I, Group D, 23
  combustible gas indicators (CGI), 32–36
  communications, 147
  electrical, 51–53, 146–147
  failures, 102
  harnesses, 170–174
  hoisting devices, 181–183
  inspection, 171f, 175
  ladders, 143, 176f
  lifting device improvisation, 143, 180–181
  loads, 166–168
  oxygen-monitoring, 36–37
  for permit-required entries, 23–24
  PPE, 67, 82, 116, 119–124
  retrieval/safety lines, 119–122, 139f–142f, 179–183
  ropes, 143–146, 183–185
  standards, 119–120, 144, 169–170, 173
  training, 185–186
  tripods, 136–138, 175–181
  wristlets, 174
ESOs (emergency services organizations). *See* Emergency services organizations (ESOs)
Evaluation
  of action plans, 135
  classification of confined spaces, 197–200
  NFPA standards, 194–195, 196–197
  NIOSH criteria, 200–201
  non-entry rescues, 197
  OSHA qualifications, 191–192, 194
  OSHA standards and, 189–192, 195–196
  response time, 189–191
Explosimeters, 32–36
Explosive range, 5, 7
Exposures, 85–86

F

FACE reports, 81
Fall protection, 181–183
FAST teams, 67
Fatalities/injuries to rescuers, 2–3
Filters, 101
Finance/administration, 72–73

Flammable atmospheres, 5–8
Flammable ranges, 5–7, 33–34
Flash points, 6, 36
Foot pounds, 167
Foot protection, 121

G

Gas-specific monitoring, 37–39, 41
Gloves, 120–121
Goals and objectives
  command and, 61, 63–65, 68–69
  confined space rescues and, 85–86, 133–134, 148
  evaluation and, 192
  incident management system (IMS) and, 59, 74
  lead time and, 88–89
Ground fault circuit interrupters (GFCIs), 147

H

Harnesses, 23f, 86f, 119–120, 170–174
Hazard and risk assessments, 49–51, 132–133
Hazardous materials, 8
Hazards. *See also* Strategic factors
  atmospheric, 4–8, 21–22, 81–83
  chemical contamination, 86, 122–124
  classification of, 195–200
  contents, 87
  engulfment, 8
  noise, 124
  physical, 8–9, 83–85
  response time and, 189–191
Hearing protection, 124
Heaters, 101, 115
Heat stress, 110–114
Helmets, 117f
Hoisting devices, 181–183
Hot work, 10
Hydrating, 112
Hydrogen sulfide, 38
Hypothermia, 114

I

IAP (incident action plans), 65, 134–135. *See also* Preplanning
Ignition sources, 82
Immediately dangerous to life and health (IDLH), 38, 62, 198
Impact loads, 166–167
Incident action plans (IAP), 65, 134–135. *See also* Preplanning
Incident command system (ICS). *See* Incident management system (IMS)
Incident management system (IMS)
  command posts, 65–66
  command structure, 68–69
  in confined space rescue, 73–75
  defined, 59–61
  incident action plans (IAP), 65, 134–135

incident commander (IC), 60–61
priorities, 67–68
public information officer, 70
resource management, 66–67, 87–88, 134
safety, 60–61, 67, 69–70, 74
single *vs.* unified command, 64
span of control, 61–62, 68
support staff, 71–73
using common terminology, 62–64
Incident stabilization, 67–68, 92
Inerting, 104
Inlet and exhaust ventilation openings, 98–101
Inspection and maintenance, 171f, 175
Interferants, 41
Internal combustion engines, 101–102

K

Kernmantle ropes, 144

L

Ladder belts, 23f
Ladders, 143, 176f
Laid ropes, 143
Latches, 51–52
Lead time, 88–89
LFL (lower flammable limit), 5–7, 32–36, 38, 200
Liaison, 71
Life safety, 60–61, 67, 69–70, 74, 89–92
Life safety rope. *See* Ropes
Lifting capacity, 176, 180–181
Limiting factors. *See* Strategic factors
Line valves, 51–52
Loads, 166–168
Location and accessibility, 83
Locking devices, 51
Lockout/tagout procedures
  devices, 51–53
  entry permits and, 22
  hazard and risk assessments, 49–51
  preplanning, 49, 53
  strategic factors, 53–54
Logarithmic scales, 39
Logistics, 72
Lower explosive limits. *See* Lower flammable limits (LFL)
Lower flammable limits (LFL), 5–7, 32–36, 38, 200

M

Maintenance and inspection, 171f, 175
Manually operated electrical circuit breakers, 51
Material Safety Data Sheets (MSDSs), 5
Mechanical ventilation, 96–98, 111–112
Medical monitoring, 113–114, 116
Methane calibration, 34
MSDSs (Material Safety Data Sheets), 5

N

National Fallen Firefighters Foundation, 108
National Institute for Occupational Safety and Health (NIOSH), 2–3, 200–201
Negative-pressure ventilation, 102–103
NFPA 1006, Standard for Rescue Technician Professional Qualifications, 20, 22, 27, 194–195
NFPA 1561, Standard on Emergency Services Incident Management System, 58, 61
NFPA 1670 Standard on Operations and Training for Technical Search and Rescue Incidents, 147–148, 150, 194–197
NFPA 1982 Personal Alert Safety System (PASS) units, 169–170
NFPA 1983 Fire Service Life Safety Rope and System Components, 184
NFPA 1983 Standard on Life Safety Rope and Equipment for Emergency Services, 119, 144, 169–170, 173
NIOSH (National Institute for Occupational Safety and Health), 2–3, 200–201
Noise, 124
Nondestructive testing, 170
Non-entry rescues, 197
Non-permit required confined spaces, 9–10, 19, 26–27
Non-sparking tools, 24

O

Objectives. *See* Goals and objectives
Occupational Safety and Health Administration (OSHA). *See* OSHA
Offensive operations, 149
Operations, offensive *vs.* defensive, 149
Operations officer, 71–72
OSHA (Occupational Safety and Health Administration)
 criteria for confined spaces, 3–4
 equipment standards, 119–120, 169–170
 Lockout/Tagout (Standard 1910.147), 48
 number of confined space entries, 15
 Permit-Required Confined Spaces (Standard 1910.146), 2, 5, 15, 20, 189, 191–192
 permit-required *vs.* non-permit required confined spaces, 9–10
 Procedures for IDLH Atmospheres 1910.134(g)(3), 62
 Procedures for interior structural firefighting 1910.134(g)(4), 62
 team roles and responsibilities, 16–19
Oxygen
 ignition sources, 82
 inerting, 104
 levels, 7–8
 meters, 36–37

P

Pacing the work, 113, 116
Parts per million (ppm), 37
PASS (NFPA 1982 Personal Alert Safety System) units, 169–170
PCBs, 86
PEL (permissible exposure limits), 22
Penetration of chemical protective clothing, 123
Pentane calibration, 34
Permeation of chemical protective clothing, 122–123
Permissible exposure limits (PEL), 22
Permit-required confined spaces
 defined, 108–109
 NIOSH criteria, 200–201
 vs. non-permit required, 9–10, 19, 26–27
 Standard 1910.146, 2, 5, 15
Personal protective equipment (PPE)
 appropriate to hazards present, 82, 85
 chemical protective clothing, 122–124
 earplugs, 124
 entry permit requirements, 20, 26
 gloves, 120–121
 harnesses, 119–120
 heat and, 109
 helmets, 117f
 mechanical ventilation and, 96
 respiratory protection, 116–118
 skin protection, 121–122
pH devices, 39–40
Physical hazards, 8–9, 83–85. *See also* Hazards
Planning chief, 71
Pneumatically powered blowers, 101–102
Positive-pressure supplied air respirators (SAR), 4, 23f, 62, 82, 117–118, 147
Positive-pressure ventilation, 102–103
PPE. *See* Personal protective equipment (PPE)
Preplanning, 49, 53, 80–81, 83–84, 196
Pressure waves, 6
Problem identification, 132
Procedures. *See* Standard operating procedures (SOPs)
Property conservation, 68, 92–93
Public information officer, 70
Pulleys, 145f, 176f, 178f

Q

Qualifications of teams, 191–192, 194

R

Radio frequency (RF) interference, 43
Rescues. *See also* Equipment; Evaluation
 action plans, 134–135
 agreements, 192–194
 attendants, 18
 basic actions, 85
 checklists, 159f–161f
 classifications of, 195–200
 command establishment, 129, 132
 communications, 22, 63–64, 89, 124, 147
 defined, 3–4, 108
 electrical equipment, 51–53, 146–147
 entrants, 18
 entry permits, 15–16, 19–26, 81
 goals and objectives, 85–86, 133–134, 148
 hazard and risk assessment, 132–133
 incident management system and, 73–75
 initial operations, 148–149
 lifting device improvisation, 143, 180–181
 monitoring, 42f
 non-entry, 197
 objectives, 133–134
 problem identification, 132
 programs for, 16–19
 resource needs, 134
 response time, 189–191
 retrieval/safety line management, 139f–142f, 179–183
 ropes, 143–146
 steps illustrated, 129f–131f
 termination phase, 135–136
 training, 147–148
 tripods, 136–138
 victims, 150–151
Resource management, 66–67, 87–88, 134
Respiratory protection
 air purifying respirators (APR), 122
 SAR, 4, 23f, 62, 82, 117–118, 147
 SCBA, 4, 62, 82, 118, 120, 147
Response time, 189–191
Retrieval equipment, 119–122, 139f–142f, 179–183
Retrieval winches, 181–182
Risk to life, 89–90
RIT teams, 67
Ropes, 143–146, 183–185

S

Saddle Vent™, 100–101, 103f
Safety
 air monitoring and, 91–92
 chemical protective clothing, 122–124
 cold-related injuries, 114–116
 incident management system and, 60–61, 67, 69–70, 74
 lines, 138, 139f–142f, 180f, 182
 medical monitoring, 113–114
 noise, 124
 officers, 61, 69–70
 overview, 108–110
 respiratory protection, 116–118

Safety (*continued*)
  retrieval equipment, 119–122, 179–183
  strategic factors, 89–92
  temperature stress, 110–114
SAR (supplied air respirators), 4, 23f, 62, 82, 117–118, 147
SCBA (self-contained breathing apparatus), 4, 62, 82, 118, 120, 147
Signage, 16
Single command, 64
Size-up, 80–81. *See also* Strategic factors
Sked™ stretchers, 144, 145f
Skin protection, 121–122
Slings, 176f
Snap hooks, 182–183
SOPs (standard operating procedures), 65
Span of control, 61–62, 68
Special problems, 91
Spectacle blinds, 51f
Spectators, 89–90
Staging areas, 67
Standard operating procedures (SOPs)
  checklists, 158, 159f–161f
  developing, 155–156
  in incident action plans (IAP), 65
  written, 156–158
Standards
  ANZI Safety Requirements for Confined Spaces (Z117.1-2003), 15, 19–20
  for equipment, 119–120, 144, 169–170, 173
  Life Safety Rope (NFPA 1983), 119, 144
  NFPA 1006, Standard for Rescue Technician Professional Qualifications, 20, 22, 27, 194–195
  NFPA 1561, Standard on Emergency Services Incident Management System, 58, 61
  NFPA 1670 Standard on Operations and Training for Technical Search and Rescue Incidents, 147–148, 150, 194–197
  NFPA 1982 Personal Alert Safety System (PASS) units, 169–170
  NFPA 1983 Fire Service Life Safety Rope and System Components, 184
  NFPA 1983 Standard on Life Safety Rope and Equipment for Emergency Services, 119, 144, 169–170, 173
  OSHA Lockout/Tagout (Standard 1910.147), 48
  OSHA Permit-Required Confined Spaces (Standard 1910.146), 2, 5, 15, 20, 189, 191–192
  OSHA Procedures for IDLH Atmospheres 1910.134(g)(3), 62
  OSHA Procedures for interior structural firefighting 1910.134(g)(4), 62
Standby assistance, 17
Static loads, 166
Static ropes, 144
Stored energy, 47–48, 54
Strategic factors. *See also* Hazards
  atmospheric hazards, 81–83
  communications, 89
  construction, 86–87
  contents, 87
  exposures, 85–86
  incident stabilization, 92
  life safety, 89–92
  lockout/tagout, 53–54
  physical hazards, 83–85
  property conservation, 92–93
  rescue size-up, 80–81
  resources, 87–88
  risk to life, 89–90
  special problems, 91
  time, 88–89
  weather conditions, 90
Stress, 110–114, 136
Supervisors, 18–19
Supplied air respirators (SAR), 4, 23f, 62, 82, 117–118, 147

T

Team evaluation. *See* Evaluation
Temperature stress, 110–114
Termination phase, 135–136, 148
Time factors, 88–89, 189–191
Training, 80, 147–148, 185–186, 192
Transformer retrieval support, 179
Tripods, 136–138, 168f, 175–181

U

UFL (upper flammable limit), 6
Underwriters Labs (UL), 119
Unified command, 64
Upper flammable limit (UFL), 6
U.S. Environmental Protection Agency (EPA), 122
Utility ropes, 144–146

V

Vapor density, 34, 35f, 37, 98
Ventilation
  equipment failures, 102
  inerting, 104
  inlet and exhaust openings, 98–101
  mechanical, 96–98, 111–112
  positive- and negative-pressure, 102–103
  power sources, 101–102
Victims, 147–148, 150–151
VOC (volatile organic compounds), 37
Volatile organic compounds (VOC), 37

W

Water-powered fans, 102
Weather conditions, 90
Wheatstone Bridge, 33
Windchill factor, 115
Working loads, 168
Wristlets, 174